D1256586

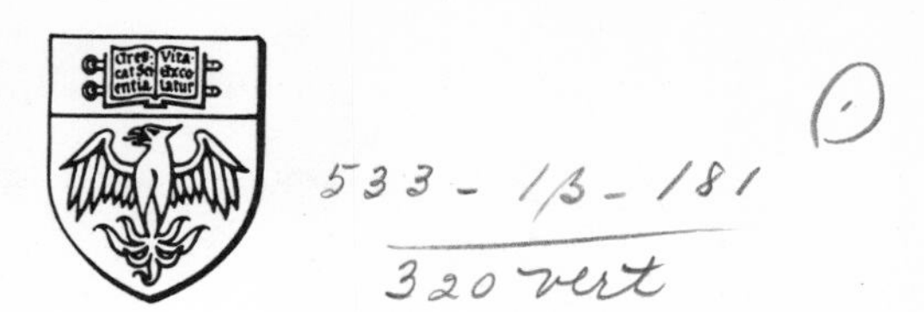

EVOLUTION OF SLOPES ON ARTIFICIAL LANDFORMS— BLAENAVON, U.K.

by

Martin J. Haigh
The University of Chicago

THE UNIVERSITY OF CHICAGO
DEPARTMENT OF GEOGRAPHY
RESEARCH PAPER NO. 183

1978

Published 1978 by The Department of Geography
The University of Chicago, Chicago, Illinois

For Kitty
and for my Parents

Library of Congress Cataloging in Publication Data

Haigh, Martin J. 1950–

Evolution of slopes on artificial landforms–Blaenavon, U.K.

(Research Paper–University of Chicago, Department of Geography; no. 183)

Bibliography: p. 271

Includes index.

1. Slopes (Physical geography)–Wales–Blaenavon. 2. Coal mines and mining –Wales–Blaenavon. 3. Anthropo-geography–Wales–Blaenavon. I. Title. II. Series: Chicago. University. Dept. of Geography research paper; no. 183.

H31.C514 no. 183 [GB448] 910s [551.4'3]
78-7267
ISBN 0-89065-090-X

Research Papers are available from:

The University of Chicago
Department of Geography
5828 S. University Avenue
Chicago, Illinois 60637
Price: $6.00 list; $5.00 series subscription

ACKNOWLEDGEMENTS

This project could not have been completed without a great deal of help from a large number of people and organizations. I am very grateful to everyone who has assisted me in this project. My especial thanks go to: Dr. G. T. Warwick, project supervisor, Miss Kitty Pearce, long-time unpaid research assistant and proof-reader, Mr. Jack Haigh, designer of equipment, collaborator, photographer, and to my Mother who has patiently put up with several years of muddied boots. I should also like to recognize the assistance given by the: Welsh National Water Development Authority (Mr. Wain), who provided the climatological data; National Coal Board (Mr. Fox), National Coal Board Opencast Executive (Mr. Hughes, Mr. Davies), and the former Monmouthshire County Council (Mr. Rodgers), who each gave permission to establish my experiments, provided maps, and supplied information concerning the area's history; also the Reprographic Section, Department of Geography, University of Keele (Mrs. Patrick, Mr. Taylor), who helped with some of the diagrams; Mr. John Hanner, who helped prepare the illustrations for publication; Dr. Warwick (University of Birmingham), Dr. E. M. Bridges (University College Swansea), and Dr. K. W. Butzer (University of Chicago), who reviewed early drafts; and Ms. Sammye Parsons, who undertook the final proof reading of this volume.

This research was undertaken as Project GT4/71/G/122 for the Natural Environment Research Council of Great Britain. Financial support for research was made available by the Natural Environment Research Council and, to a lesser extent by the University of Keele. Financial support for the publication of this monograph has been provided by the Department of Geography, University of Chicago.

Martin J. Haigh
Chicago 1977

CONTENTS

LIST OF TABLES

LIST OF FIGURES

Figure		Page

Figure | Page

Figure | Page

LIST OF PHOTOGRAPHS

CHAPTER 1

THE STUDY OF MAN-MADE LANDFORMS

There are very few parts of the world which continue to possess a wholly natural landscape. Man's influence on the face of the earth is pervasive and increasing dramatically. Most modern landscapes are full of landforms which, if they do not directly or indirectly result from man's activities, have, at least, been substantially modified by his works.

It is unfortunate that geoscientists have been slow to recognize this situation. The traditional reaction of geomorphologists has always been to "take to the hills" and to seek out "natural landscapes" where human activities and human influences are minimal (vide: Beaumont - 1973), but how wise is it to try to ignore the geomorphology of man? Certainly, with regard to physical geography, the neglect of the human angle has had unhappy consequences--emphasizing the schism between "human" and "physical" geography, encouraging questions concerning the "relevance" of geomorphology, and re-inforcing the arguments for linking geomorphology with geology (vide: Butzer - 1973). Simply, the failure to tackle the study of the man-made landscape amounts to a retreat from reality. The facts of the matter have been repeatedly stated: Golomb to the I. G. U. in 1964: "In scale and intensity, human landform modification begins to rival that due to 'natural' processes"; Zapletal (1973c): "Nowadays, the most energetic exogenetic agency is man." Ryabchikov (1971) calculates that 85% of the Earth's dryland surface has been substantially modified, that 55% is intensively employed, and 3% completely destroyed by man.

Studies of the physical geography of the artificial landscape are generally considered under the heading of "Anthropogenic Geomorphology." Most modern studies of anthropogenic geomorphology trace their origins back to the works of the American, George Perkins Marsh (1864, 1877) who wrote much of the fundamental catechism of the study: "Man is everywhere a disturbing agent. Whereever he plants his foot, the harmonies of nature are turned to discords. The proportions and accommodations which insured the stability of existing arrangements are overthrown . . ." (Marsh - 1877, pp. 131-32). "Man extends his actions over vast spaces, his revolutions are swift and radical and his devastations are, for an almost incalculable time . . . irreparable" (Marsh - 1877, p. 133). Other studies attribute the creation of their discipline to the writings of another author--R. L. Sherlock (1922, 1923, 1931). Sherlock's discussions of the volumes of excavated materials, of land subsidence and the spread of mining waste reflect continuing themes in the literature of anthropogenic and environmental geomorphology. "Man's geological agencies are primarily as an agent of Denudation . . . In addition, in mining operations he disturbs the flow of underground waters and caused subsidences . . . Man disturbs the courses of rivers: fills lakes and makes new ones, checks or promotes sea-erosion and modifies climates . . . His great engineering feats are so widespread that it would be a herculean task to sum up their total effect on Nature" (Sherlock - 1931 pp. 8-10). This is perhaps the reason why Sherlock confined his studied to his native Britain.

Unfortunately, neither Marsh's nor Sherlock's studies made any lasting impression on the geomorphological literature of the English language world, though both continued to receive attention elsewhere (Popov - 1968; Legget - 1969 Zapletal - 1968, 1973b). Recently, the rise of Environmental Geomorphology in the United States has revived interest in Marsh (Coates - 1972; Lowenthal - 1974). Nevertheless, the development of anthropogenic geomorphology has largely been due to the work of German authors like Edwin Fels (1934, 1935, 1955, 1965). It was Fels who first coined the phrase "Anthropogenic Geomorphology" and who first recognized the fundamental division of anthropogenic landforms and processes into those which are "Direct," or consciously created, and those which are "Indirect," or unconsciously created. Another major contribution was made by Russian authors such as Bondarčuk (1949) and Devdarijani (1954) who was responsible for the concept of the "wild-becoming" of man-made landforms. However, it was the Czechoslovakian, Ladislaw Zapletal, who formalized the subject with this definition. Anthropogenic geomorphology is the "geographical characterization and geomorphological classification of man-made and man-induced landforms" and the study of "the processes by which anthropogenic landforms arise, develop, and become extinct" (Zapletal - 1973b).

Recently, there have appeared several articles which attempt to define the scope of anthropogenic geomorphology by considering the ways in which man effects processes which influence landform generation. Tables 1.1 and 1.2 are two attempts to accomplish this task. The first (Table 1.1), which is from Demek (1973), presents an over-view of the theoretical potential of the subject. The second (Table 1.2), which is based loosely on the writings of Zapletal (1968,

TABLE 1.1

MAIN MODES OF MAN'S EFFECT ON THE EARTH'S RELIEF
(Demek - 1973)

1. Affecting endogenic processes
   - producing artificial earthquakes
   - producing isostatic movements, mainly surface subsidences

2. Affecting exogenic processes

2.1 acceleration of exogenic processes, especially
   gravity processes
   fluvial processes
   thermokarst processes
   abrasion processes

2.2 retardation of exogenic processes, especially
   fluvial processes

3. Effects of anthropogenic processes
   - anthropogenic degradation, such as
   - lowering, planation
   - subsidence
   - excavation
   - anthropogenic aggradation
   - anthropogenic transportation

1973a, b), Cooke and Doornkamp (1974), and others, is designed to describe the main-stream of thought and practice within the subject at the present day.

Figure 1.1 introduces another dimension of the discipline. This is Iogansen's classification of anthropogenic landscapes, a classification based on an analysis of the different grades, qualities and intensities of human interference (Iogansen - 1970). Its great merit is that its divisions may be correlated with the history of landscape alteration from the relatively minor disturbances of early man through to the progressively greater alterations occasioned by the agricultural and industrial revolutions. The scheme also encompasses the most recent trends in anthropogenic landform generation by recognizing the importance of reclamation and the future possibility of the evolution of planned landscapes "where natural environmental linkages are scientifically controlled in the interests of the whole society" (Iogansen - 1970).

This brief review of the study of artificial landforms is concluded with Table 1.3. This classification of directly created anthropogenic landforms, which was devised by Zapletal (1968, 1969), may be regarded as a crude stock list of the main classes of man-made landform. It may be significant for the rest of this volume that the first two categories of this scheme

TABLE 1.2

CLASSIFICATION OF ANTHROPOGENIC LANDFORMING PROCESSES

1. Direct anthropogenic processes

1.1 constructional
- tipping: loose, compacted, molten
- graded, molded, ploughed, terraced

1.2 excavational
- digging, cutting, mining, blasting of cohesive or non-cohesive materials
- cratered
- trampled, churned

1.3 hydrological interference
- flooding, damming, canal construction
- dredging, channel modification
- draining
- coastal protection

2. Indirect anthropogenic processes

2.1 acceleration of erosion and sedimentation
- agricultural activity and clearances of vegetation
- engineering, especially road construction and urbanization
- incidental modifications of hydrological regime

2.2 subsidence: collapse, settling
- mining
- hydraulic
- thermokarst

2.3 slope failure: landslide, flow, accelerated creep
- loading
- undercutting
- shaking
- lubrication

2.4 earthquake generation
- loading (reservoirs)
- lubrication (fault plane)

refer to landforms created by mining and industrial surface forms such as waste dumps.

Anthropogenic geomorphology is a subject with a respectable pedigree, a substantial level of theoretical development, and a vast importance. Its origins lie in the writings of the Western World, but it has matured mainly in Eastern Europe. This is a strange situation but there may be several explanations. One explanation might be framed by a fashionable invocation of the "poverty of Western geomorphology" during the first half of this century. There is no doubt that this was a period when academic study found more favor than applied

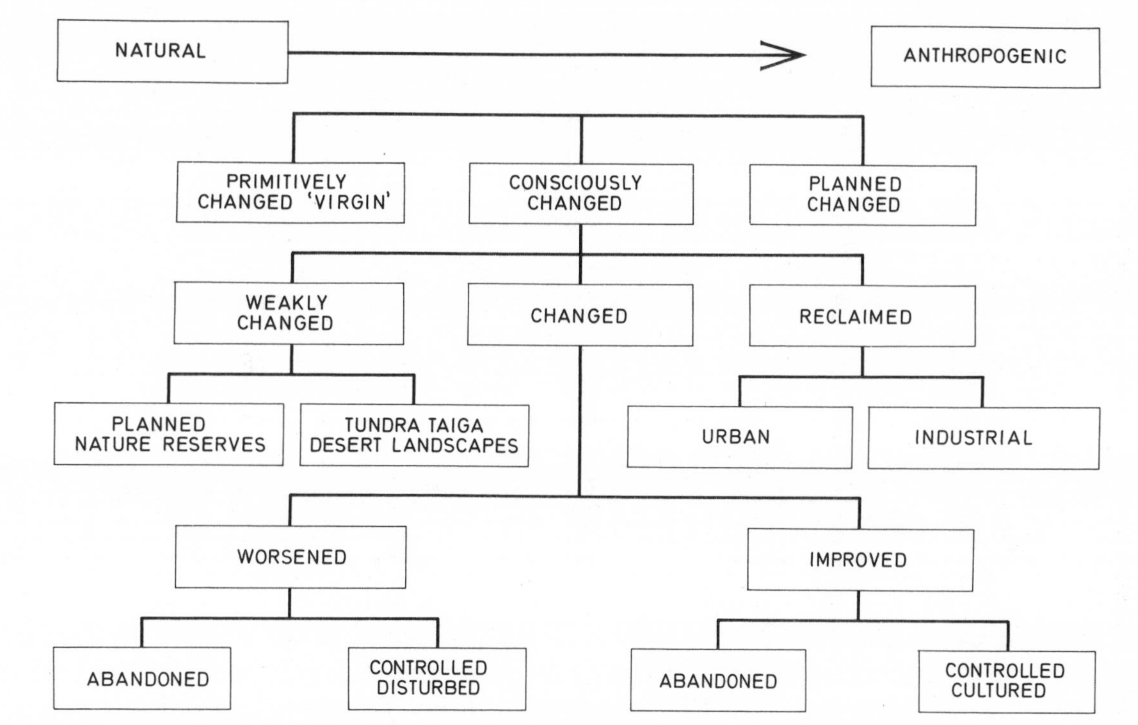

Fig. 1.1.--Classification of Anthropogenic Landscapes (Iogansen - 1970)

TABLE 1.3

MORPHO-GENETIC CLASSIFICATION OF
ANTHROPOGENIC LANDFORMS
(Zapletal - 1968)

1. Landforms due to Mining Activity
   including: spoil mounds, loose and compacted terraces, subsidences, mining excavations
2. Landforms due to Industrial Activity
   including: spoil mounds--loose, compacted and molten tipped, storage dumps of fuels and raw materials, landfill plateaux and terraces
3. Agricultural Landforms
   including: agricultural terraces, planations, drainage dykes and ditches
4. Urban Landforms
   including: city mounds and earthworks, landfill plateaux, hydraulic subsidences, and river channel modification
5. Landforms due to Communications
   including: tunnels, embankments, canals and river channel modifications
6. Coastal Landforms
   including: breakwaters, artificial beaches and groynes
7. Military Landforms
   including: craters, earthworks and moats
8. Mounds created as Memorials

research, and when academic geomorphology was strongly dominated by the more esoteric implications of the Davisian Cycle and the study of Historical Geomorphology. This may have been an environment which was far less suitable to the development of the subject than that provided by the utilitarian attitudes enforced across Eastern Europe after the Second World War. The perpetuation of the subject's isolation may also be given a sociological explanation. The "Iron Curtain" is not just a political, cultural, and economic barrier. It is also a major barrier to information diffusion (Haigh - 1975). The existence of this barrier has prevented awareness of the development of this subject from penetrating the Western academic consciousness.

This study does not comprise the first crusade for the recognition of anthropogenic geomorphology. There were several similar sermons presented to the I.G.U. Conferences of the 1960s (Barthel _et al_. - 1964; Golomb - 1964; Zapletal - 1960, 1964). Unfortunately, these articles made little impact. It is probable that there still exists among academic geomorphologists a perpetuation of the old idea that "man-made landforms don't count." It is to be hoped that the 1970s, a period when the problems of the environment have become an established vogue and when academic research is increasingly being asked for its own justification, will prove a more receptive era for the growth of world-wide studies of man-made geomorphology.

CHAPTER 2

INTRODUCTION

This study is a contribution to the field of Anthropogenic Geomorphology. Its special concern is the evolution of slopes on directly created constructional landforms which have appeared in the course of the exploitation of the coal resources of South Wales. These landforms may broadly be considered as having one of two origins: either they are the results of the disposal of waste materials produced during deep mining operations, or they reflect the results of an attempt to replace a disturbed land-surface after the closure of an open pit operation.

The unnatural shapes which are the result of the disposal of deep-mine coal spoil are common features of the landscape of South Wales. Several types of tip may be recognized, each a reflection of a particular waste-disposal technology, sometimes modified by the local relief of its site and variations in the nature of the waste materials. Tandy's illustrations of the main categories of deep-mine spoil mound encountered in the South Wales area are included as Figure 2.1 (Jones et al. - 1972). This study is concerned with three of these types: (1) conical tips resulting from MacClane tipping, (2) high plateau mounds, which are not, in this instance, topped with cones, (3) low multiple fan ridges.

The landforms which result from opencast mining tend to be less obtrusive, at least when viewed from the ground surface, but they cover far greater areas. This study is concerned with the morphogenetic evolution of two rather atypical opencast sites. The sites are atypical because one was never reclaimed and the other's reclamation failed. Both sites have since reverted to the sort of "badlands" appearance which was described by Schumm (1956a) in his classic study of the infilled clay-pit at Perth Amboy, New Jersey. The sites do, however, reflect the major division in any geomorphological typification of infilled open-cast sites, which is that one is substantially independent of the surrounding landscape while the other is functionally just a small part of a larger land system and open to imports of water and sediment.

Today, many of the old forms of deep-mine spoil dump and many of the

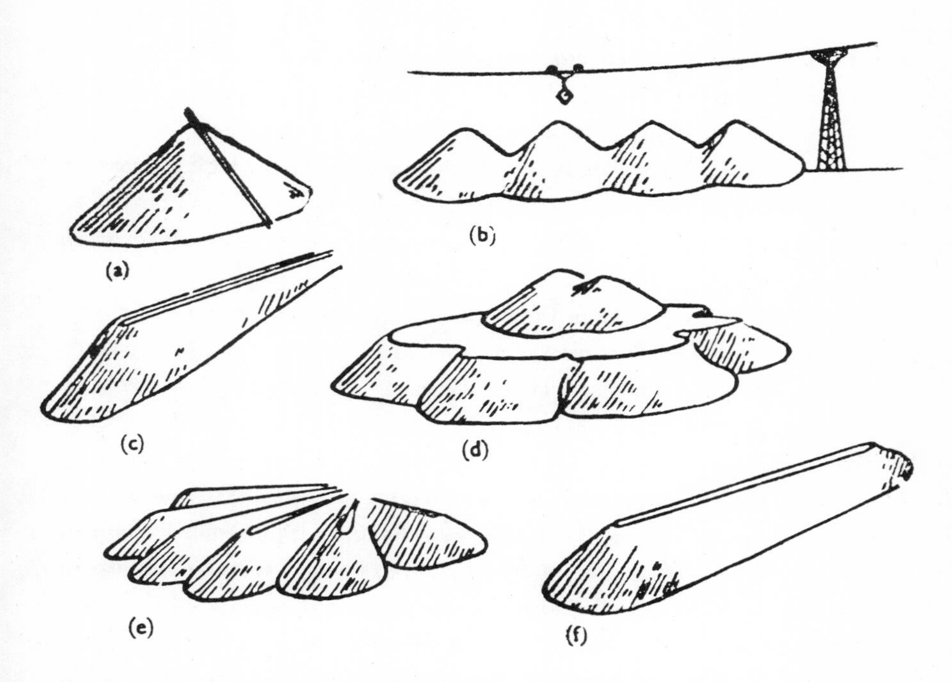

Fig. 2.1.--Un-natural Shapes of Shale Tipping (after Jones et al. - 1972)

(a) Conical resulting from MacClane tipping
(b) Multiple cones tipped from aerial ropeways
(c) High 'fan-ridges" by tramway tipping over slopes
(d) High plateau mounds topped with cones
(e) Low multiple "fan-ridges" by tramway tipping
(f) Lower ridge by tramway tipping

old opencast spoil banks are being re-shaped by land reclamation projects. These new landforms are composed of the same materials as the old spoil dumps, are often constructed in a similar fashion, and possess a final morphology which may not be essentially very different from the landforms which they replace. Reclaimed spoil mounds have not been tackled by this study, but it is suggested that their future evolution and the present-day evolution of certain of the older spoil banks described by this study, potentially, have a great deal in common.

The setting for this study is the old industrial district at Blaenavon. The artificial landscape of Blaenavon is treated as a natural laboratory for the comparative study of the morphogenetic development of critical sites on several species of spoil bank.

The artificial landscape of Blaenavon is also utilized for the establishment of some related experiments which may have a wider theoretical significance. The Blaenavon landscape contains a large number of essentially simple, reasonably datable, and relatively comparable landforms, many of which have plausible and important analogues in the "natural" landscape. These landforms are evolving from an approximately known original state with the unusual rapidity that is characteristic of young landforms which are out of equilibrium with their environment. Their rates of evolution are, therefore, relatively easy to measure and the results of such experiments easy to interpret.

It is a contention of this study that the materials and construction of these spoil banks are not wholly singular. The artificial slopes constructed by bulldozer from the clayey overburden of the two opencast sites are not greatly different from slopes developing anywhere on recently dumped, exposed, or devegetated clays (cf. Schumm - 1956a). Similarly, the free-standing scree slopes of the MacClane Shute cone tips are not very different from the slopes of volcanic cinder cones (cf. Mizutani - 1969, 1970).

The principle of using artificial landforms for the study of the evolution of natural hillslopes has been validated by the publication of the geomorphological results from the British Association's experimental earthwork at Overton Down (Crabtree - 1972; Jewell - 1963), and the reports of hundreds of laboratory studies which, by the use of flume or rainfall simulator, attempt to discover natural relationships.

In this study, experiments have been established which attempt to discover the effects of aspect, vegetation, and gullying on the rates and pattern of slope development. Since the climate and processes which affect these experiments can only be those which also affect all sections of the natural landscape, it is to be expected that the morphogenetic responses to these influences must also, in part, echo those which would have been recorded in a study of part of the natural landscape.

Philosophically, the case for regarding artificial landform evolution as a model for natural landform evolution is quite straightforward. Both systems exist in the same environment and, within the confines of these experiments, both are affected by the same stimuli. Any theory purporting to explain landform evolution in the one situation would therefore necessarily possess the same calculus as one purporting to explain the other. The causal sequences invoked to explain slope evolution in both situations would, thus, both be interpretations of the same calculus and would contain identical theoretical structures. Events which would follow as logical consequences in the natural system would inevitably also follow in the artificial system. According to Braithwaite (1953), the main condition for the acceptance of an analogue of a theory as a model for that theory is that it should have the same calculus and the same theoretical structure.

However, it must always be remembered that a model is not the theory itself. Braithwaite (1953) notes that although one may construct models for atomic behavior by discussing atoms as "solar systems" of separate elementary particles, solar systems possess attributes other than those which make the model appropriate. Thus, although it is possible to argue that studies of the development of artificial landforms may be used to aid the understanding of theories for the evolution of natural hillslopes, it must always be remembered that artificial landforms possess attributes which have no place in the natural system. Consequently it must be considered dangerous to assume that theories induced for the evolution of artificial hillslopes would _necessarily_ prove appropriate to the examination of natural hillslope evolution, though, obviously, in many cases they must.

CHAPTER 3

A HISTORY OF ANTHROPOGENIC LANDFORM GENERATION IN THE BLAENAVON AREA

Blaenavon is a striking example of an area whose landscape is completely dominated by artificial landforms. The economic incentive for the creation of most of these features was the presence of large reserves of good steam coal. However, coal mining was not the original, nor is it the current, stimulus for anthropogenic landform generation.

The earliest anthropogenic landforms in the area were created by iron-stone working. Surficial deposits were the first to be extracted and the primitive patch-work/scouring method was adopted. Small streams were ponded, then flooded into the miner's hollows. The surge flushed out the lighter shales and left behind the heavier ironstones. Patch-working of the Blaenavon iron-stones began around 1584 at Elgam Hill (S0256097), and it caused a storm of protest from local farmers who found that their meadows became fouled by grey shale outwash (Jones - 1974). Patch-working entailed: ". . . virtually turning over the mountainside like a garden" (Lloyd - 1906), and it remained the dominant mining technology in the area for at least a century. The Elgam Hill area is now the site of a council housing estate but considerable tracts may still be discovered in the Blaenavon area which possess the characteristic hummocky aspect of a patch-worked landscape. Today, these low rounded irregular mounds are overgrown by heath and moorland grasses. The local name for this landscape is, appropriately enough, "Tumble."

Inevitably, the surface deposits of iron-stone eventually became worked out. Patch-working, therefore, gave way to underground mining. Little documentary evidence exists of these early iron-stone drifts and levels. The main memorial to this activity was the creation of numerous small "fan-ridge" spoil mounds which are intermingled amongst those due to contemporary coal mining activity.

Coal mining began at Blaenavon in 1782. The first colliery was called "Bridge Level" and was sunk near the present Dragon Site Reclamation. The

P. 3. 1. --Garn Pit (1902) on the Site Now Occupied by the Waunaton Opencast

expansion of the coal mining industry in the area was rapid. In 1906, an eighty year old miner, L. Browning, was able to list by name over thirty local collieries. These old collieries were excavated underground by the old bridge and stall method. The morphology of numerous local subsidences reflect these underground structures, but the most obtrusive reminders left from the early coal workings are the large numbers of low, flat-topped, fan-ridge coal tips.

Morphologically, colliery spoil mounds can be divided into three types and each type can be related to a particular tipping method (Denison - 1959).

(I) High Ridge Tips: formed by tipping from an aerial bucketway or from an extending lattice girder frame resting on the actual tip.

(II) Conical Tips: created by a conveyor belt system such as a MacClane Shute which is laid on one face of the tip.

(III) Flat Topped Tips: formed by tipping rail wagons or rubber tired dumper trucks operating over the flat surface of the tip and advancing on one or more faces.

Most of the tips in the Blaenavon area belong to the third category and within this class it is possible to recognize two types:

(a) Fan Ridge Tips: these are the most primitive. They were created by tipping from tramways extending from the entrance of the drift or level. When the tipping front moved too far from the source of the waste the tramroad was moved. This process built up a tip complex with a characteristic radiating plan (vide: Figure 3.1). Fan Ridge Tips were constructed extensively in the Blaenavon area throughout the nineteenth century. However, construction seems to have faded out with the start of the twentieth century.

(b) Plateau Tips: these belong mainly to the first part of the twentieth century. They were created by a dual process of tipping and compaction. They are characterized by very steep, almost recti-linear slopes, a lobate plan, and a relatively great local relief. Sometimes, on the steeper valley sides, these tips are dumped as distinctive layered terraces (vide: Figure 3.2, Photograph P.3.2).

These flat topped tips are the main anthropogenic landforms created in Blaenavon in the nineteenth and early twentieth centuries, and, together with the early iron-stone scourings, they affected an area of around 1,000 acres.

The most obtrusive anthropogenic landforms ever created at Blaenavon were, without a doubt, the MacClane Shute spoil mounds, which were constructed from the early part of the century up until the mid-1960s. These conical tips

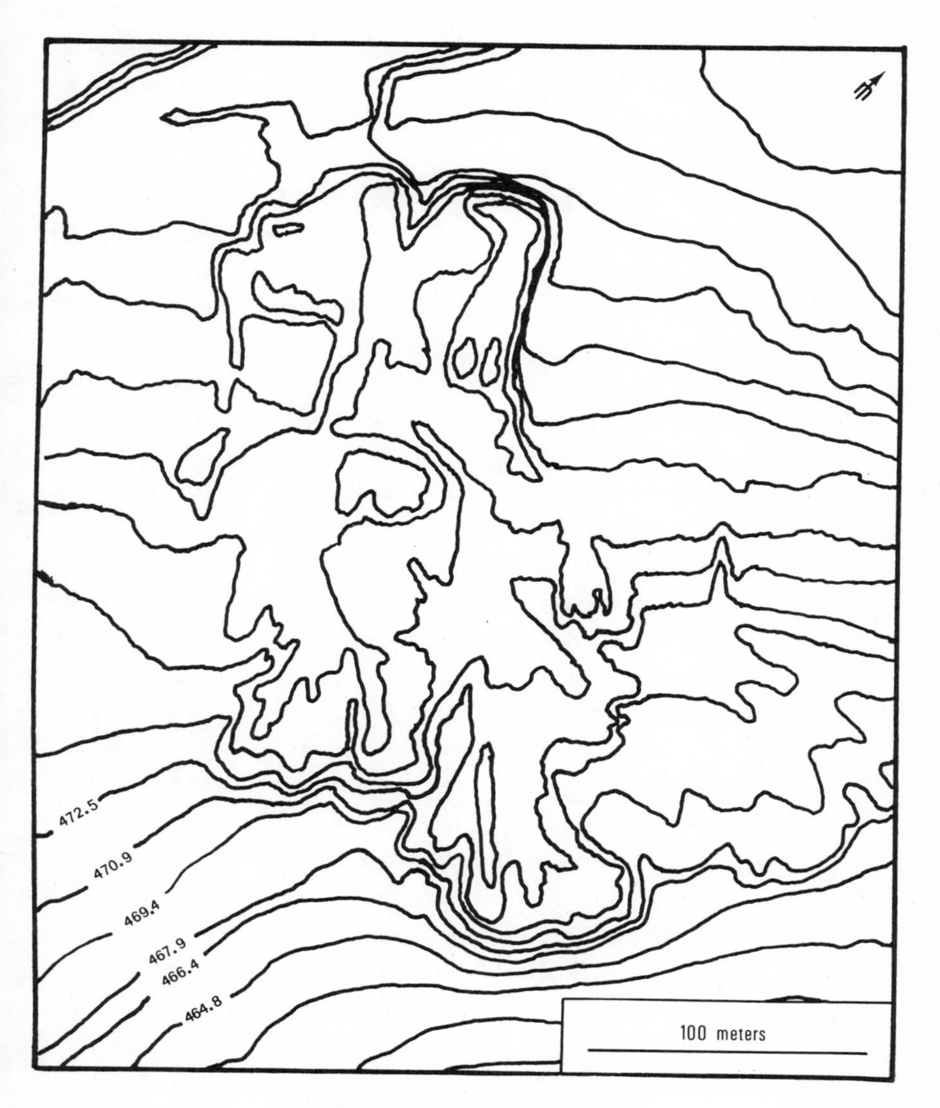

Fig. 3.1.--Fan Ridge Spoil Tips

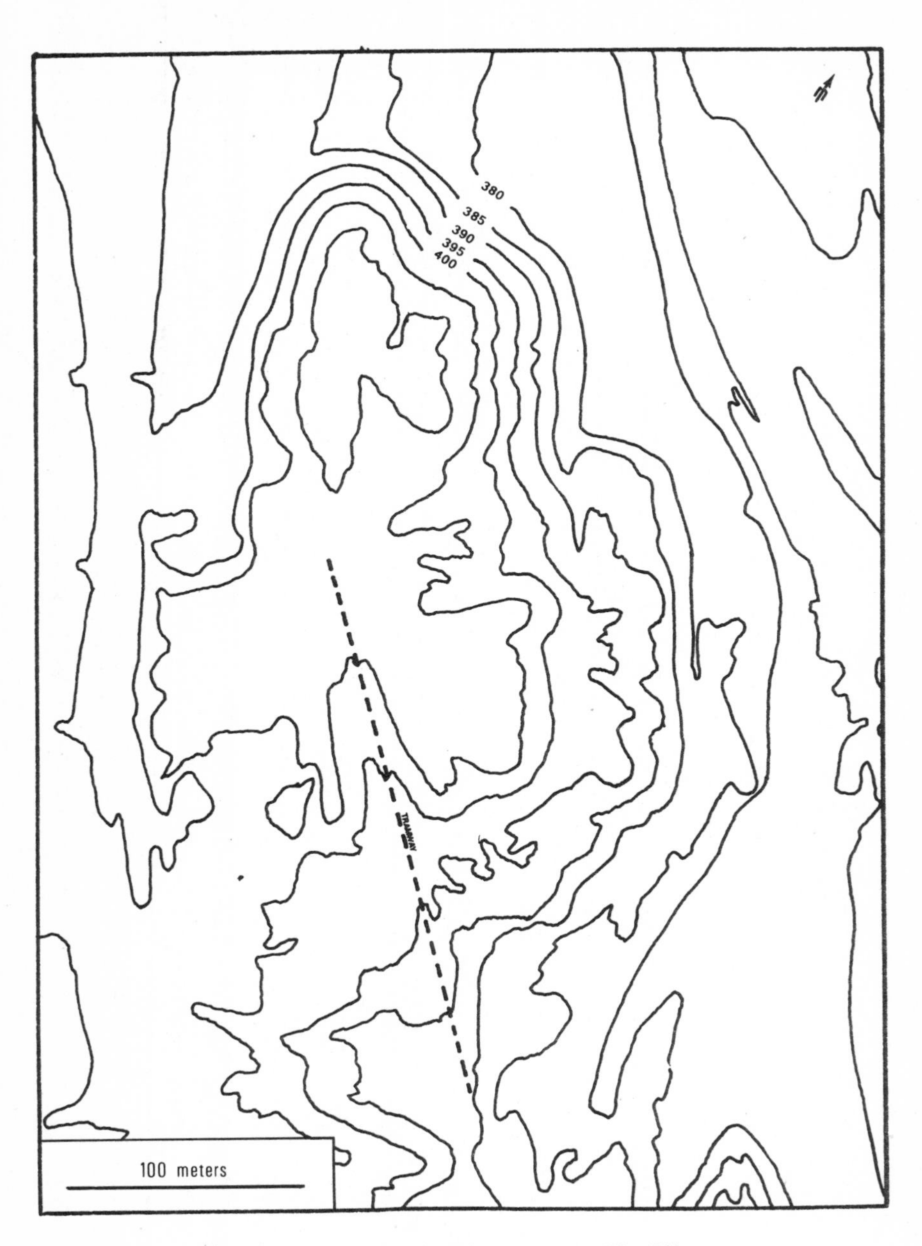

Fig. 3. 2. --Plateau Spoil Tips near Big Pit

P. 3.2. --The Big Pit Area: Plateau spoil tips and MacClane Shute Cones 1972

contain up to several million tons of spoil between them, and attain heights in excess of 50 meters. Characteristically, they are pear-shaped in plan, and wedge-shaped in profile. Their shape was controlled by the extension of the MacClane Shute which formed the hypotenuse of the wedge, and the long axis of the pear-shaped plan. The tips' free slopes were controlled solely by the spoil's angle of rest, and tended to be steepest where the tip was youngest and tallest, beneath the mouth of the MacClane Shute. The slope of the spine of the tip is controlled by the angle at which the shute was established. This method of tipping gave rise to very unstable, scree-like, cones, which were largely unvegetated and gullied (Figure 3.3).

Three of the four Blaenavon MacClane Shute cones are typical of their type. The fourth is rather special. This is the Washeries Tip (Figure 3.4, Photograph P.3.3), now one of the largest tips surviving in South Wales, and composed entirely of very fine coal washings. Washeries waste is fairly cohesive and has the capacity to form a crust some centimeters thick. It is entirely devoid of vegetation.

There is relatively little evidence of the spontaneous combustion of the Blaenavon spoil mounds. Some of the plateau tips have suffered burning, and one of the MacClane Shute cones, that at Keare's Level, is said to have lost 10m in altitude through combustion. Burnt spoil is pink in color, it is soft, friable, and relatively fertile.

The Second World War instigated a new phase of mining activity in the Blaenavon area. Opencast mining began in 1942 using imported American machinery. The largest of these new sites was opened up by MacAlpines at Waunafon and it operated until about 1948. Two smaller pits were opened up around the same time at Pwll Du, and two deep "exploration" trenches at Blaen Pig presumably date from the same period. Figure 3.5 is a plan of one of these large excavations.

The main opencast pits were infilled when they were exhausted. There are indications that some rough attempt was made to resculpture the land surface and to restore former drainage courses like that of Cefn Garn-yr-erw on Waunafon (Figure 3.6). However, the official view of today's N.C.B. Opencast Executive is that these sites are not "reclaimed." This would seem to be a fair statement. The sites, which are surrounded by a 10-20m scarp, are intensely gullied and poorly vegetated (Photographs P.3.4 and P.3.5).

After the closure of the Waunafon Opencast, the same machinery was transferred to excavate the site at Waun Hoscyn. This site stretches three kilo-

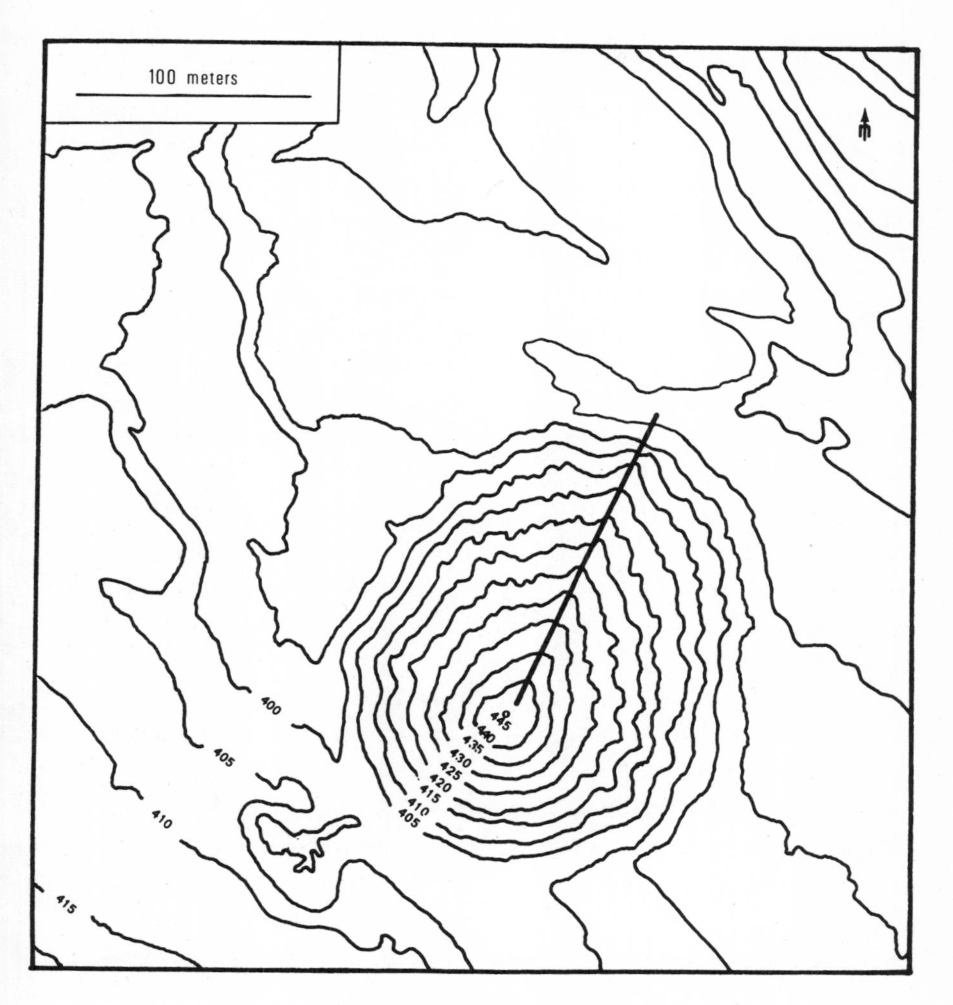

Fig. 3. 3. --MacClane Cone near Big Pit

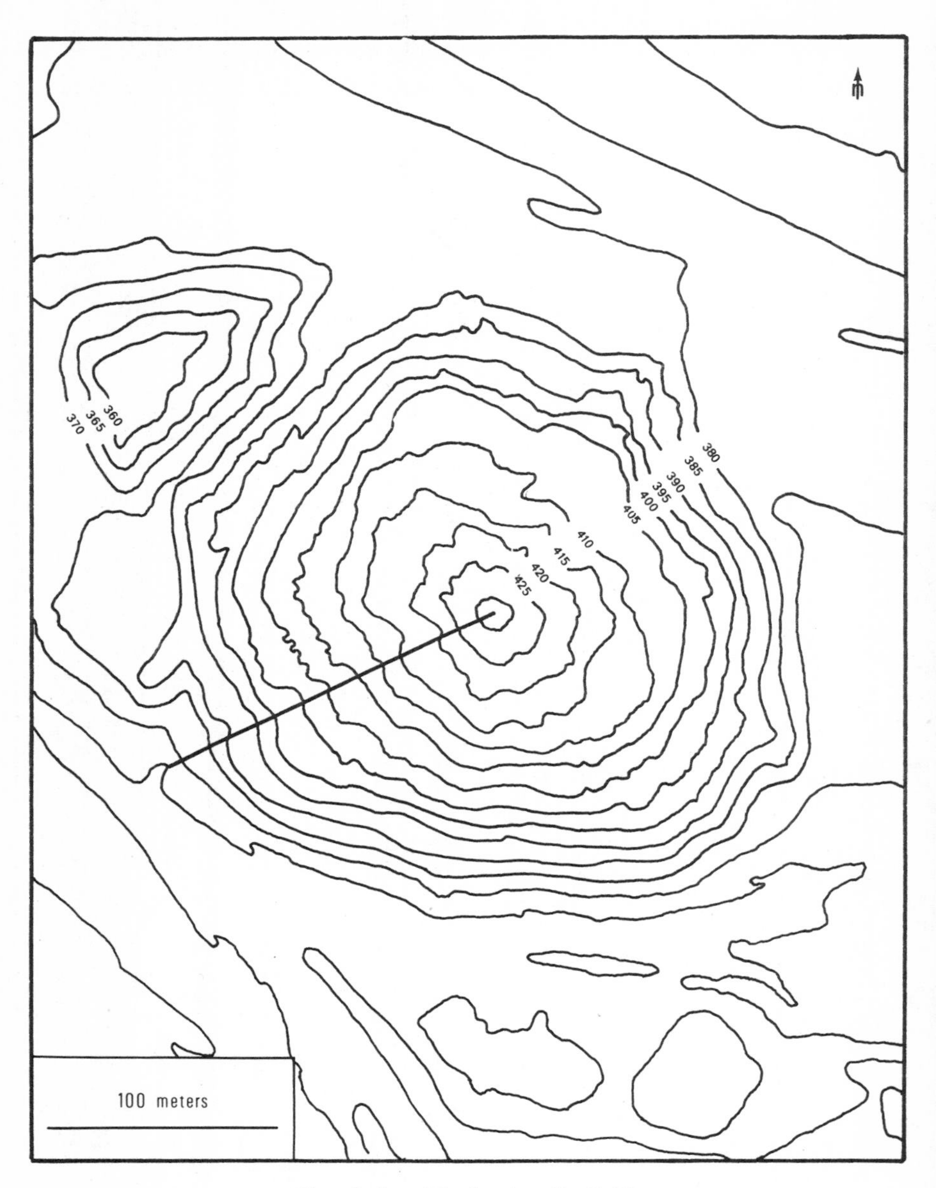

Fig. 3.4.--Washeries Spoil Tip

P. 3. 3. --Washeries Spoil Tip with Snow in Gully Channels

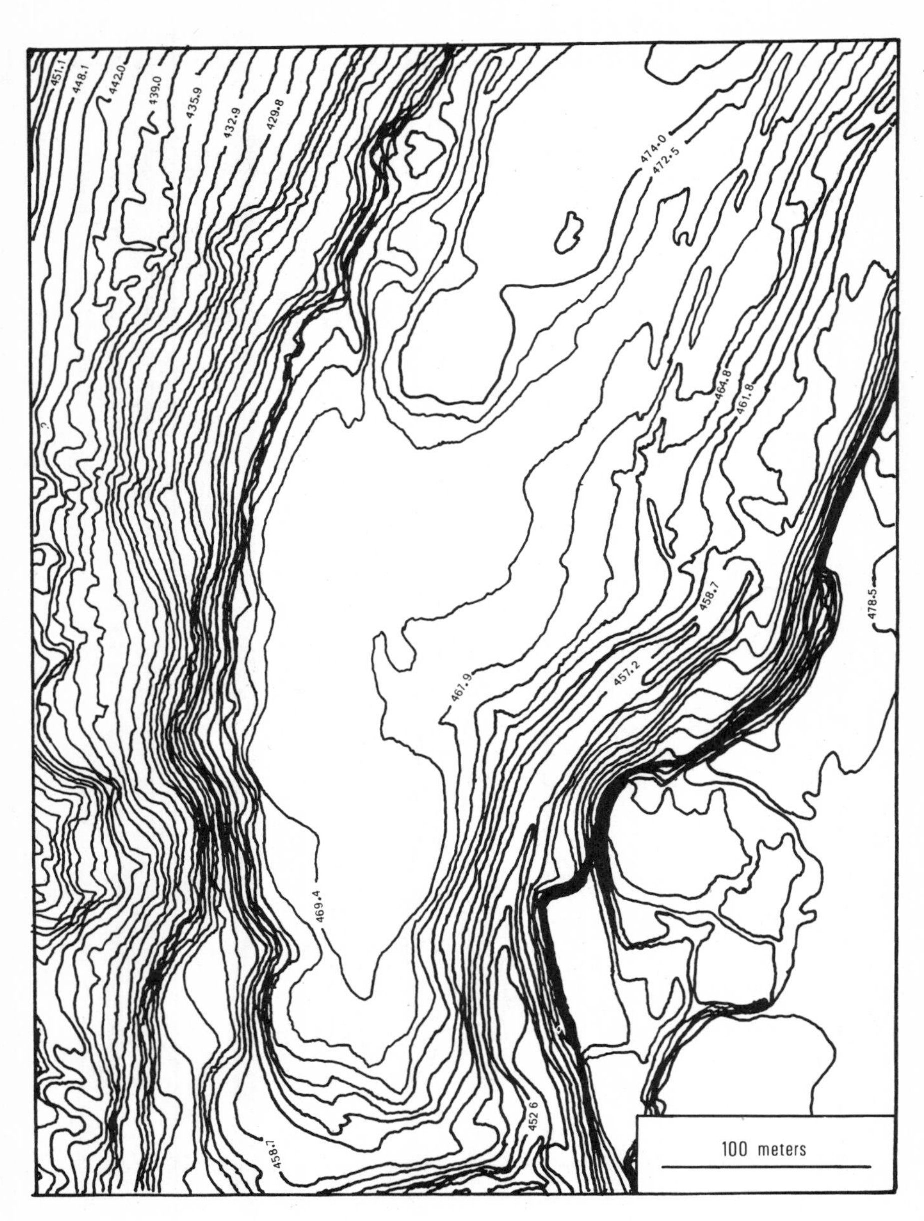

Fig. 3.5.--The Blaen Pig Trench

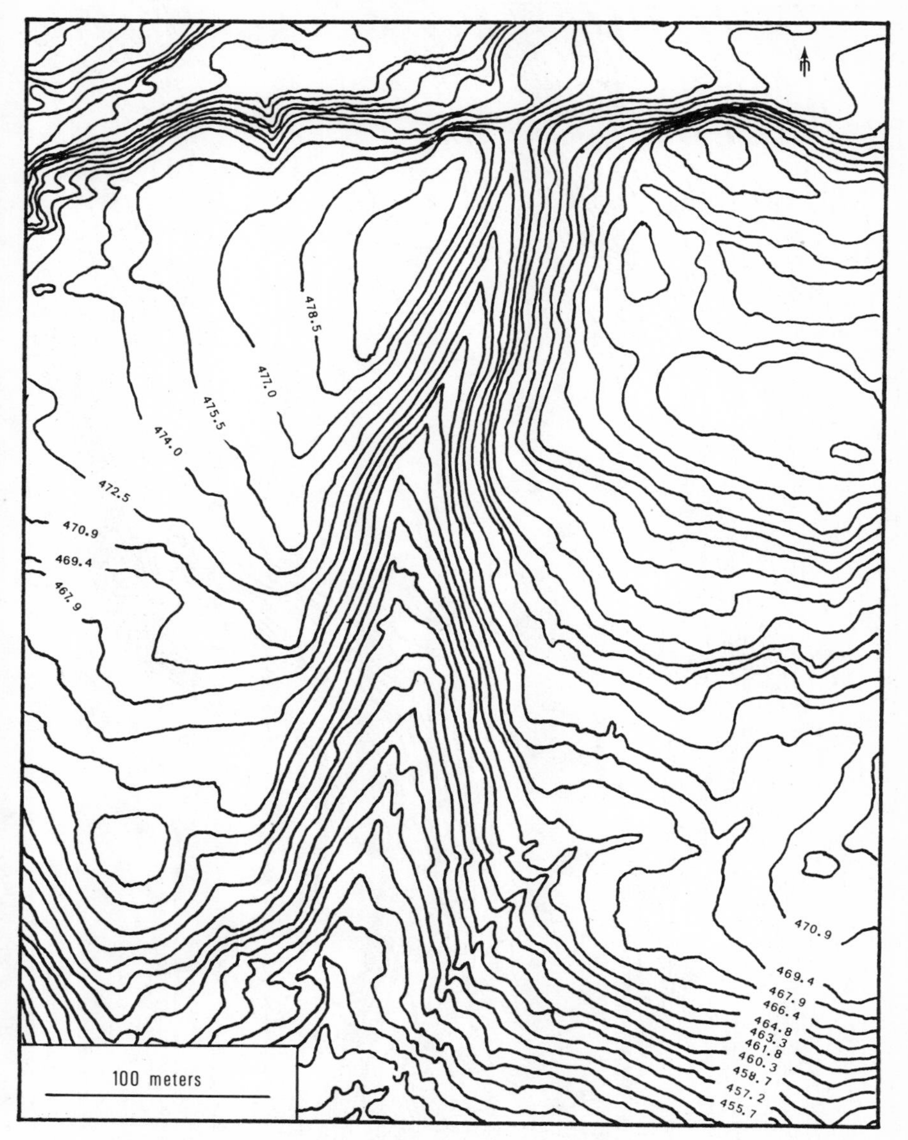

Fig. 3.6.--Cefn Garn-yr-erw: A restored drainage course on the Waunafon Opencast fill.

P. 3. 4. --Scarp Edge of the Waunafon Opencast Fill with Deep-mine Fan Ridge Spoil Dumps in the Foreground.

P. 3. 5. --Unvegetated and Gullied Spoil of the Eastern Waunafon Opencast

meters along a high valley bench to the immediate south-west of the town of Blaenavon. The Waun Hoscyn site was restored as rough mountain grazing by its coaling contractor in 1956. However, this restoration was unsuccessful, vegetation growth was poor and gullying was severe. Subsequently, the site was selected by the N. C. B. Opencast Executive for the experimental comparison of conventional and hydromatic seeding methods. These experiments occupy a 55-acre site as 11 strips of 33m. The experiments were established in June-August 1970 and seem to be enjoying considerable success.

The closure of the Waun Hoscyn Opencast began a sixteen year lull in opencast activity in the Blaenavon area. This was ended in the early 1970s when work began to open up a new site at Llanelly Hill adjacent to the old Waunafon Opencast and using the same site for its screenings, the Lion Disposal Unit. The new site was opened up in 1972 and rumors of its impending closure became current in 1975.

However, it is unlikely that the closure of the Llanelly Hill site will mark the end of opencast activity in the area. According to the report of the Monmouthshire Derelict Land Reclamation Joint Committee (1971), opencast operations are scheduled for 650 acres of land east of the Garn Road, and adjacent to the Waunafon and Pwll Du sites.

The most recent phase of anthropogenic landform generation in the Blaenavon area is both the most extensive and ambitious to date, and, so it is to be hoped, the most lasting. The Industrial Development Act of 1966, and the Local Government Act of 1966, made land reclamation a viable proposition for local authorities in the national Development Areas by providing 85% grants. In 1971, the M. D. L. R. J. C. published its "Comprehensive Report on the Extent of Derelict Land" which contained proposals for the reclamation of 595 acres of derelict land in the area. It is possible that over half of this objective will have been achieved by 1975.

Land reclamation schemes give rise to a distinctive breed of landform which most closely resemble the flat topped tips. They are formed by tipping, compaction, and grading. They tend to have smooth recti-linear slopes and are characterized by a plentiful provision of land drains, grass seedings and, locally, tree plantings. Their slopes tend to be less steep than those of most plateau spoil tips and they tend to be designed to provide the greatest possible flat plateau top. Figure 3. 7 is a plan of one of the earlier reclamations, Dragon Site, and illustrates both typical reclamation morphology, and the way in which the older spoil tip contours are resculptured.

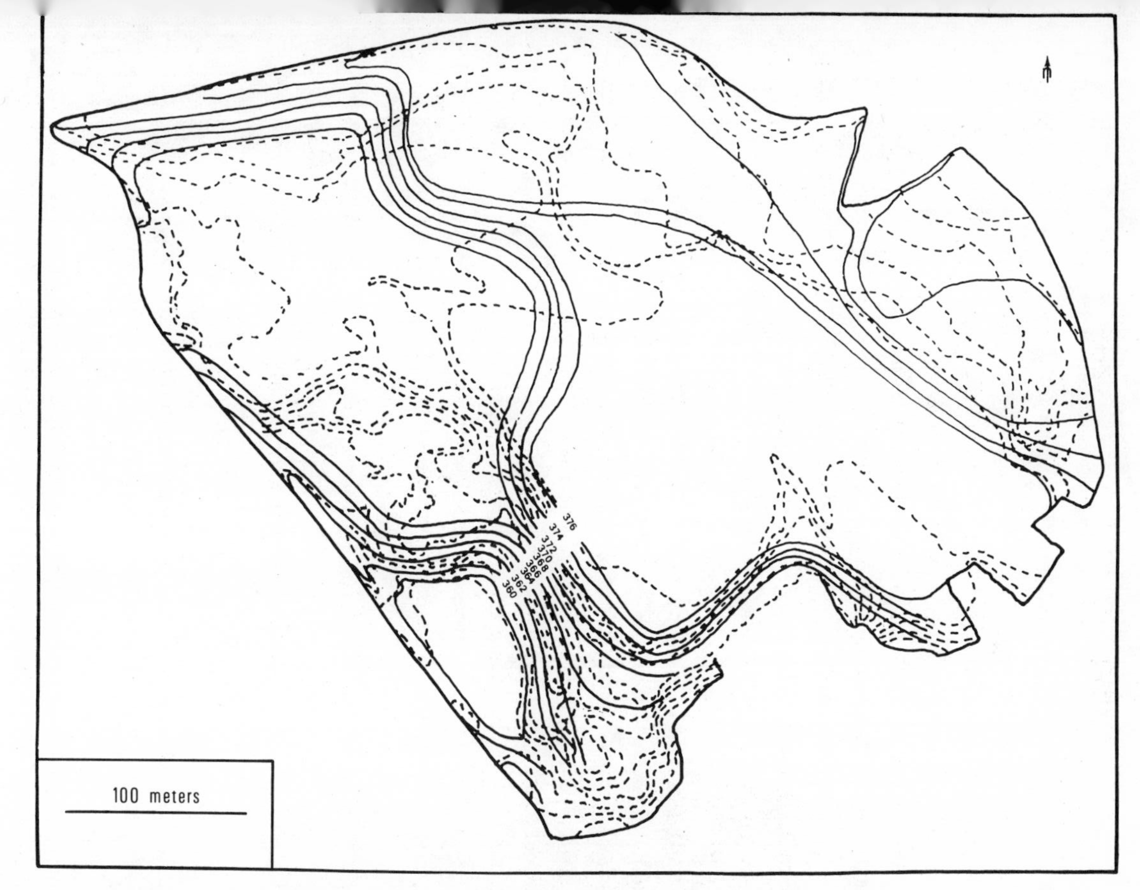

Fig. 3.7.--The Alteration of Spoil Contours by the Land Reclamation Process: Dragon Site. (Former contours - dashed line; current contours - solid line.)

P. 3. 6. --Cliffs of Contorted Iron Slag near the River Arch in 1972

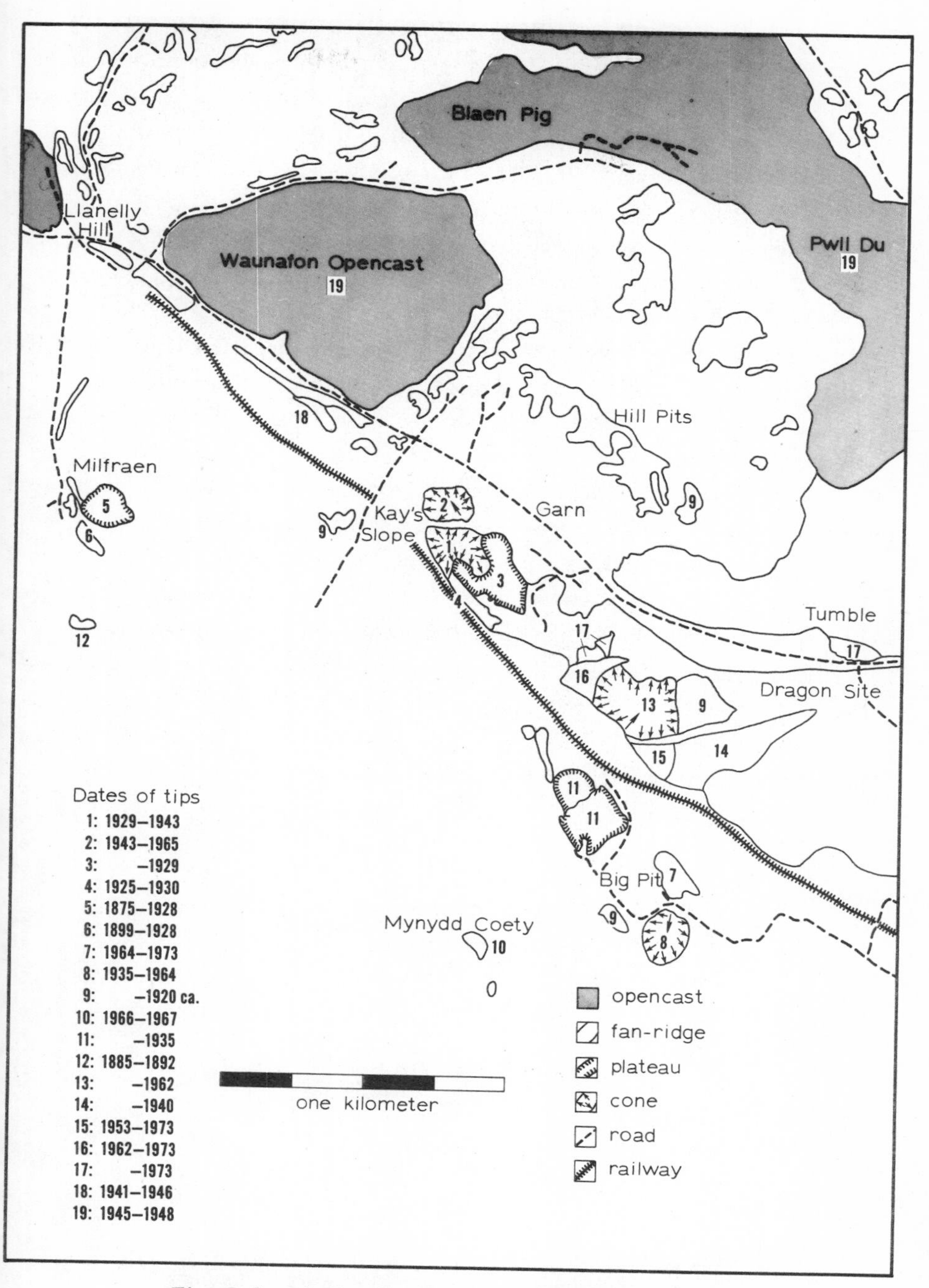

Fig. 3. 8. --Anthropogenic Landforms at Blaenavon

At the time of writing, reclamation mainly affects the area between the Washeries Tip and Blaenavon. Virtually all of the old spoil contours from the Garn road in the north-east to Waun Hoscyn in the south-west have been altered The MacClane Shute Tip at Big Pit was levelled in Spring 1973. The cliffs of contorted slag (Photograph P. 3. 6), a relic of Blaenavon's iron and steel industry, which formerly dominated the approach to the River Arch, where the Afon Lwyd reappears from beneath coal spoil, were also destroyed at this time. The next land-mark to disappear will probably be the MacClane Shute Cone at Kay's Slope. This was scheduled for reworking by Ryan's Holdings Ltd., who would have spread its waste across some of the more devastated sections of the Waunafon Opencast. The site's current status is, however, unclear.

In the artificial world of the future, students of landscape will find themselves drawn increasingly to the consideration of artificial landforms and deposits. Blaenavon is one of the oldest industrial areas of South Wales, and as such presents a scale of anthropogenic landscape interference which must become increasingly common in years to come. The physical landscape of Blaenavon (Figure 3. 8) is dominated by the legacies of past economic exploitation, by present planned redevelopment, and by contemporary engineering technology.

This chapter is concluded by a sequence of photographs depicting stages in the technogenetic modification of the Blaenavon landscape by land reclamation. The final photograph of this sequence, taken in 1974, depicts the nascent Gilchrist-Thomas industrial estate. This estate is the trump card of the planners' plans for the revitalization of the Blaenavon economy.

P. 3. 7. --The Washeries and the Regrading of the Slag Cliffs, 1972

P. 3. 8. --Regrading at the River Arch and Big Pit's Tip after Reclamation, 1973

P. 3. 9. --Land Reclamation at the Gilchrist-Thomas Industrial Estate, 1973-1974.

CHAPTER 4

THE EVOLUTION OF SLOPES (REVIEW)

"Slope studies have always been among the most perplexing which have faced the geomorphologist . . ." (Chorley - 1964, p. 70). Problems arise because, potentially, there are a great many complicating factors which may affect the process of slope evolution. These complicating factors include: lithology, structure, tectonic movements, climatic change, changes in the vegetation cover, and the impact of man. Frequently, the precise nature of these changes and their impact on slope morphology is unknown (Schumm and Moseley 1973).

The long histories and complicated structures of most natural slopes have lent them a complicated and essentially varied morphology which is inevitably very difficult to interpret. These problems are enhanced because, in the natural world, present forms can be interpreted as being the consequence of one or several past or present factors, and complicated forms may be generalized in any of several ways to conform with a variety of preconceived notions.

A consequence of this is that the study of slopes has become a good example of a branch of geomorphology which has become top-heavy with theory. The subject suffers from a plethora of contradictory slope models based, all too frequently, on a maximum of verbal, or more recently mathematical, speculation (Davis - 1898, 1899; Penck - 1924 [1953]; Scheidegger - 1961ab; Hirano - 1968; Ahnert - 1971; Luke - 1972), and a minimum of precise field measurement--which is expensive, time-consuming, and very boring. It is very difficult, thus, to test the validity of such models in the field except in very special circumstances (Fair - 1947, 1948; Savigear - 1952; Schumm - 1956b) or for relatively small areas (Ahnert - 1970). It is much easier merely to engage in verbal debate about the validity of one or other of the slope models, but it should always be remembered that a theory is only as strong as its weakest premise and that such a weakness need not necessarily be exposed by deductive logic. Further, there is always the danger that discussion of such models will divert attention away from reality (Yatsu - 1966; Butzer - 1973). Chorley (1964) notes that one ". . . impediment facing slope studies is the doctrinaire attitude of most researchers . . . resulting from strongly held cyclic, polycyclic, posthumous climatic, climatic morphometric, or for that matter dynamic/equilibrium notions . . ." (Chorley - 1964, p. 70). A further impediment is attributed to preconceived notions concerning "the importance . . . [and] complexity of present processes."

It is the intention of this study to follow a lead set by two recent reviews of slope theory (Carson and Kirkby - 1973; Young - 1972). Slope evolution is considered, as far as is possible, in the absence of the complications afforded by geology, climatic change, historical circumstance, and classical theory. The process of slope alteration is examined, first, as a suite of morphological alternatives and, latterly, interpreted in terms of observed and measurable controls of debris transportation and removal.

(I) Slope Alteration Terminology

Morphological alteration can be expressed in terms of the geometric changes which affect particular slope units. These slope units are of two types, either they are recti-linear--in which case they are referred to as slope segments, or they are curvi-linear--in which case they are called slope elements (Young - 1964). The geometric changes which affect slope segments may be reduced to parallel advance or retreat and positive or negative slope decline. Changes affecting slope elements are related to positive or negative variations in the radius of curvature. The extent of a slope unit is defined by its morphology and by its mode of geometric alteration. This adds further possibilities for slope alteration by the extension, replacement, subdivision or amalgamation of slope units. A detailed terminology has been developed by Young (1964, 1972) from Savigear (1952, 1956) for the description of individual slope units and this is displayed as Figure 4.1.

There are four conditions of debris removal and transportation which are implied by particular patterns of geometric alteration. Slope units may be variously labelled as: (1) denudation units if there is net ground retreat, (2) accumulation units if there is net ground advance, (3) transportation units if the unit is active but there is no net advance or retreat, (4) relict units if there is no debris transport and no significant advance or retreat. Both transportation slope units and relict slope units may be described as being morphologically neutral.

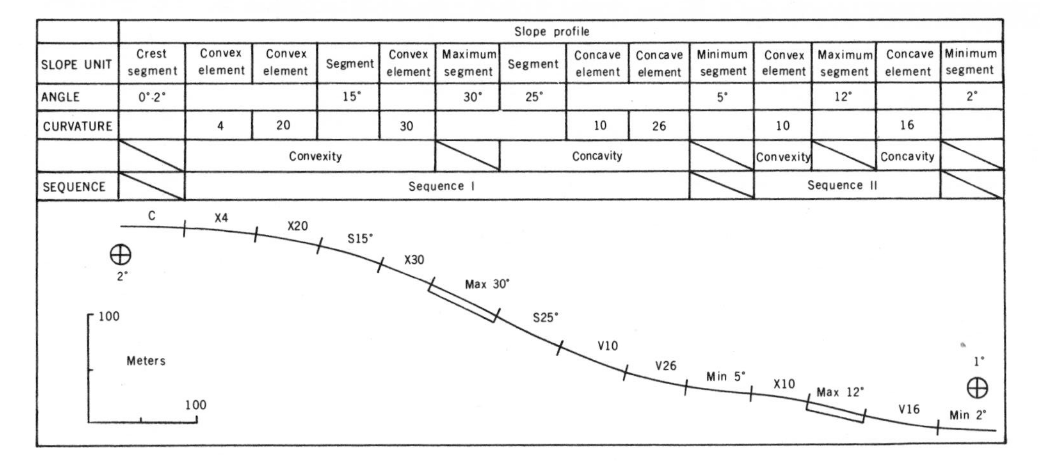

	Slope profile													
SLOPE UNIT	Crest segment	Convex element	Convex element	Segment	Convex element	Maximum segment	Segment	Concave element	Concave element	Minimum segment	Convex element	Maximum segment	Concave element	Minimum segment
ANGLE	0°-2°			15°		30°	25°			5°		12°		2°
CURVATURE		4	20		30			10	26		10		16	
		Convexity					Concavity				Convexity		Concavity	
SEQUENCE		Sequence I									Sequence II			

Fig. 4.1.--Terminology of Slope Profile Analysis. (Values of curvature are given in degrees per 100 meters.) (After Young - 1972, 1964.)

(II) Models of Main-slope Alteration

The most obtrusive portion of most slope profiles is the steep section which separates the upper convexity and lower concavity. This section, which may include, or even correspond to, the maximum slope segment is referred to as the main-slope. When erecting hypotheses to predict or describe slope evolution under various conditions of denudation and accumulation, it is often easiest to couch these hypotheses in terms of the behavior of the main slope. Figure 4.2 depicts Carson's views of the three main modes of main-slope alteration (Carson and Kirkby - 1973).

Basically, these hypotheses, when considered in terms of a simple five unit hill-slope consisting of horizontal crest segment, horizontal basal segment, steep main-slope segment, convex upper element, and concave lower element, describe the following alterations. "Main-slope retreat" (Carson and Kirkby - 1973), requires the parallel retreat of the main-slope segment at the expense of an abbreviation of the crest segment and resulting in the development and extension of the lower element. "Main-slope decline," as illustrated, is accomplished wholly at the expense of the crest segment. "Main-slope shortening" is accomplished by an extension of the upper element at the expense of both main-slope and crest segments.

A noticeable feature of these alternative modes of slope evolution is that, apart from the crest segment which is usually treated as being relict, virtually all the slopes considered are denudation slopes. A weakness, therefore, of all such systems, as they stand, might be that they all require a net removal of debris from the system without considering the conditions of that removal.

(III) Influence of Basal Control

Obviously, the condition of removal which is of paramount importance to all slope evolution is the nature and activity of the basal control. If there is no basal removal of detrital material, an inevitable result of the retreat of any part of the slope which is not effected by deflation, settling, solution etc., will be accumulation at the slope foot. The classic case of such a pattern of slope evolution is that experienced by an abandoned cliff or quarry section (Figure 4.3, and cf. the theoretical studies of Fisher - 1866; Lehmann - 1933; Bakker and Le Heux - 1947, and school).

The most usual basal control of slope evolution is a stream channel. It is not the only control however, since there are many slopes whose basal control is the sea, a lake or reservoir, a debris flow or landslide, a glacier or ice-

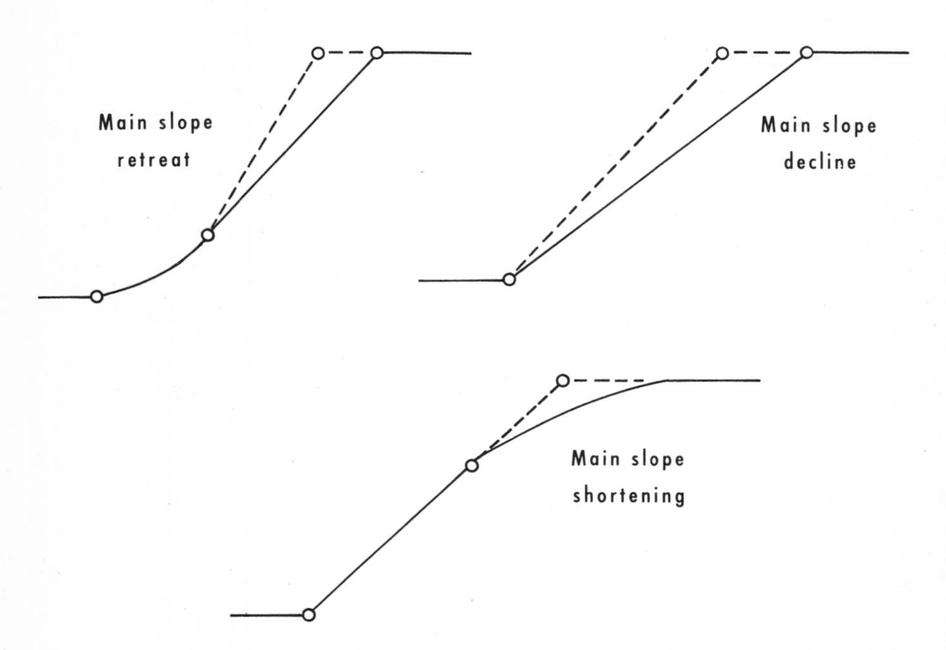

Fig. 4.2.--Three Modes of Main-slope Behavior (after Carson and Kirkby - 1973).

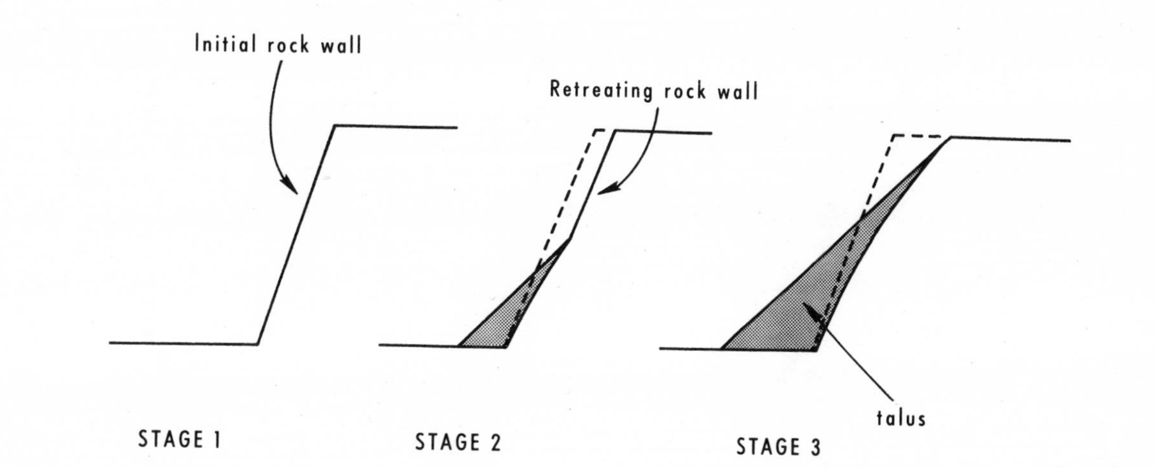

Fig. 4.3.--Theoretical Cliff and Talus Sequence Assuming No Removal of Talus from the Base. (After Carson and Kirkby - 1973.)

sheet, a foot-path or roadway, or another slope.

The activities of the basal control can be considered under two headings according to their effect upon the slope's local relief and the slope's effective length. The effect of a basal control is nearly always positive. (There are, however, some exceptions. Consider the situation of a slope whose basal control is another slope in a ditch section with no effective lateral transport. In such a circumstance, the deposits from one slope might effectively reduce both the relief and effective length of the opposed slope.) The positive effects of the basal control begin with the restriction of slope foot accumulations and extend, via dissection, to increasing the relief available for slope development (Figure 4.4). Both processes tend to favor main-slope steepening and extension.

Movements of the basal control are also important. Migrations of the stream channel, thus, can variously extend the distance available for slope development or restrict it by undercutting the base of the slope. Kennedy's statistical studies of valley sections in five areas of North America have demonstrated that the most powerful control of local variations in valley slope angle is the location of the stream channel vis à vis the slope foot (Kennedy - 1973; see also: Melton - 1957).

Variations in the basal control through time also have an important effect on slope evolution. Just as the progressive rejuvenation of a whole valley system results in the generation of a series of knick-points which migrate upstream, so too, smaller scale variations in basal control may be transmitted up-slope as travelling inflexions--bands of relative erosion, deposition, or neutrality (cf. Pinczes - 1972; Marosi and Szilard - 1968; Marosi - 1972). The importance of these effects depends on the violence and time-scale of the variations involved. An extreme case of such a process is reported by Daukza and Kortarba (1972) who describe the upslope generation of land-slides on basally undercut mountain slopes in the Carpathian mountains.

(IV) Influence of Initial Morphology

Finally, the mode of slope evolution is important because the shape of the slope at any given time is of itself an important control of the nature of its future evolution. Lixandru (1968) has demonstrated that there is a critical slope length for the initiation of erosion, which on 7° slope plots established in a Romanian orchard was 0.35m. The effects of slope length and degree on ground retreat are also well established. It has been demonstrated that, in the absence of any detailed considerations of soil unit-weight (Foster and Martin -

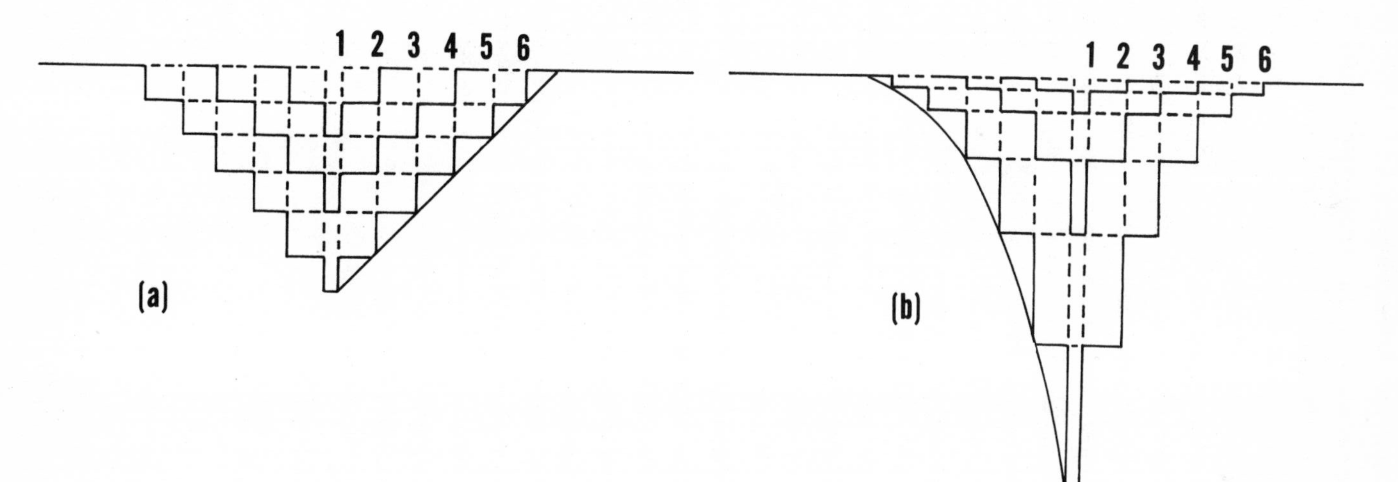

Fig. 4.4.--Slope Development under a Steady Rate of Rockfall and (a) a steady rate of stream-lowering and (b) an accelerating rate of stream downcutting (after Carson and Kirkby - 1973).

1969; Rowlinson and Martin - 1971), ground retreat increases exponentially with the slope length and slope angle. Zingg (1940) found that soil loss (tons/acre) was proportional to the 1.49 power of the slope angle (%) and 0.53 power of slope length (ft.). Musgrave (1947) found that soil loss (tons/acre) was proportional to the 1.35 power slope angle (%) and the 0.37 power of slope length (ft.). Smith and Wischmeier (1957) describe a more complicated relationship:

$$A = 0.43 + 0.30\ S + 0.043\ S^2$$

where A is the loss of soil (tons/acre) and S is the slope (%). Soil loss (tons/acre) was found to vary as the 0.5 +/- 0.1 power of slope length (ft.). These last two quotients may be combined to produce a single equation to describe the effect of both slope angle and slope length:

$$LS = \frac{L^{\frac{1}{2}}}{100} (0.76 + 0.53\ S + 0.076\ S^2)$$

where LS is the dimensionless topographic factor of the "Universal Soil Loss Equation" (Wischmeier and Smith - 1960), L is slope length (ft.), and S is slope angle (%). These results are applicable only to recti-linear segments but demonstrate that increases in the length and declivity of a slope unit tend to cause increases in its rate of retreat.

More recently, studies have been directed to a consideration of the effects of different slope unit curvatures on the rates of erosion. Young and Mutchler (1967, 1969ab) discovered that retreat and runoff varied significantly on slopes of similar lengths, and similar mean angles but with different surface configurations and that these differences were essentially independent of surface cover. Maximum soil displacement was found to occur on the upper third of a 23-meter plot with accumulations developing at the slope foot. The maximum soil displacement on recti-linear and convex slopes occurred about three quarters of the way down the slope unit. Meyer and Kramer (1968, 1969) have adopted a computer simulation approach to extend the results of Zingg (1940) and Smith and Wischmeier (1957) to an analysis of the effect of slope shape on ground retreat. The equations employed were:

(i) Zingg (1940)

$$\text{Sediment Load} = 5.0 \times 10^{-6}\ S^{1.4}\ L^{1.5}$$

(ii) Smith and Wischmeier (1957)

$$\text{Sediment Load} = 1.585 \times 10^{-5}\ (0.43 + 0.30\ S + 0.43\ S^2)\ L^{1.5}$$

(iii) Smith and Wischmeier with the addition of critical limit criteria (cf. Lixandru - 1968).

$$\text{Sediment Load} = 1.32 \times 10^{-5}\ L^{1.5}\ (S - Sc)^{1.4}$$

where S was the slope angle (%), L was the slope length (ft.) and Sc was $50/L^{0.5}$, and each computer iteration was set to equal the average annual soil loss on a Southern Indiana silt loam cropped to continuous corn. The results of this study are illustrated as Figure 4.5. It was found that fifty periods of erosion changed the profile of the convex shape most and the concave slope the least. There was a tendency for both the recti-linear and convex slopes to develop concave profiles. After 200 periods of erosion, all the slope shapes examined had developed concave profiles. These results are simply the logical consequences of the empirical erosion equations of Zingg (1940) and Smith and Wischmeier (1957) applied, out of context, to isolated non-linear slope units. They are relatively easily explained. The upper section of the convex slope is most gentle and develops least runoff, further down slope both runoff and slope angle increases and this affects a relatively large soil loss. Concave slopes have their greatest steepness where there is least runoff and hence suffer a smaller soil loss. The suggestion made is that, in the long term, and provided the assumptions made by Zingg (1940), Smith and Wischmeier (1957), and Meyer and Kramer (1968, 1969) are correct, then all agricultural and artificial slopes should evolve towards domination by their lower concavities.

Now, the above analyses refer only to sheet-flood erosion. It has been well established that the domination of a slope by different processes may give rise to slopes of different morphology. Schumm (1956b) has demonstrated that in the badland environments of South Dakota, the dominance of creep over rainwash erosion causes the development of broadly rounded divides and convex slopes, but where the processes of surface-wash predominate then steep, straight, parallel-retreating slopes tend to develop (vide: also Young - 1974). Clayton and Tinker (1971) found from their studies in the same area that, in practice, 99% of soil removal was accomplished by surface-wash, concentrated and unconcentrated. Nevertheless, the possibility of the influences of soil creep, enhanced in this instance by the local lithology, together with the discrepancy of the observed and expected morphological response to surface runoff, should both be considered as qualifications to the considerations outlined in the previous paragraphs.

It is, of course, possible that these discrepancies may be removed by the consideration of a new variable: the soil unit-weight. In 1969, Foster and Martin reported the results of a laboratory investigation into the relationship between slope angle, unit-weight (i.e., compaction) of soil, and erosion. Their analyses indicated that, "for a given unit-weight, there is a unique slope from which the maximum amount of erosion will occur" (Foster and Martin - 1969,

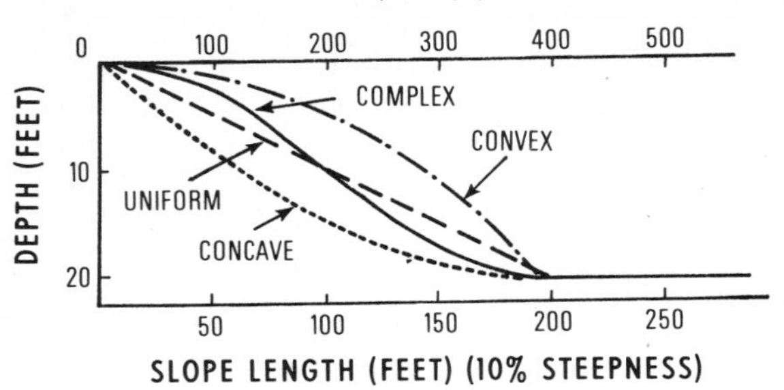

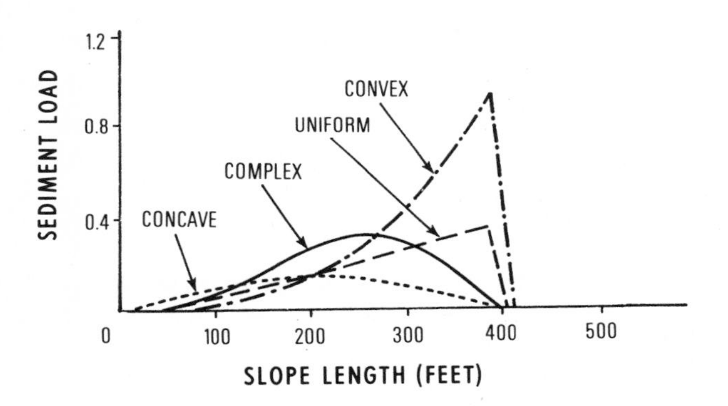

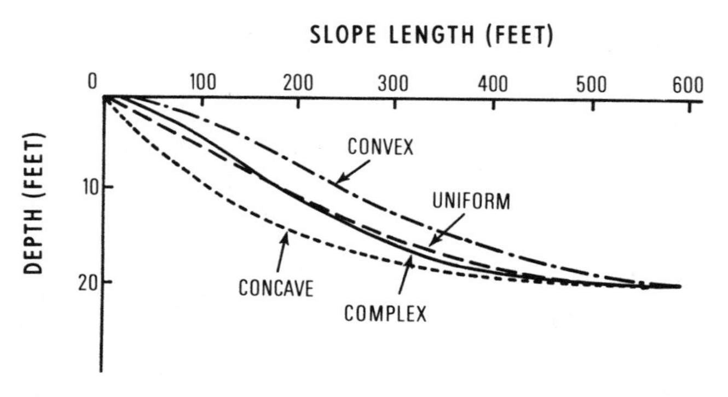

Fig. 4.5.--Effect of Slope Shape on Surface Erosion (after Meyer and Kramer - 1969).

p. 560). The erosion rate was greatest for low unit weights on flatter slopes and greatest for high unit weights on steeper slopes. Further study by Rowlinson (1968) and Rowlinson and Martin (1971) revealed that the results suggested by the studies of Zingg (1940) through to Meyer and Kramer (1969) may have a basic flaw in that they have oversimplified a rather complicated relationship. Rowlinson and Martin (1971) attempt to describe these rather complicated relationships existing between the factors which control erosion and they produce evidence that there exists a surface of maximum erosion rate which is a function of slope, unit-weight, and the depth of water-flow over the soil. This surface, which is not completely described, is controlled "in part by the detachment of soil particles due to impact and in part by the transporting capacity of both the runoff water and raindrop impact" (Rowlinson and Martin - 1971, p. 50). It is possible that the theory of unit-weight will eventually provide the link between the observed patterns of erosion recorded by Zingg (1940), Young and Mutchler (1969ab) et alia which are theoretically examined by Meyer and Kramer (1968, 1969), and the patterns of slope development which are observed in the field (Schumm - 1956b). However, this problem has not been overcome by research at the present time and there is little evidence that the importance of the work on unit-weight and its effect on erosion has yet been widely appreciated.

However, even in the absence of further research in this direction it is still possible to follow certain of the suggestions contained by the studies of the "Unit-weight" school by a more detailed consideration of the nature of the erosion processes and their effects on slope form.

(V) Microprocess and Slope Morphology

Fundamentally, slope erosion is effected by two process systems. The first process sequence begins with the addition of water as rainfall and continues as the water moves through and across the soil surface towards the ground water and local fluvial system. The second process system is a result of the natural vibrations of the soil which are due to heating and cooling, freeze-thaw, and wetting and drying. The two systems are, of course, to some extent interactive.

(a) Rainsplash

The erosional effect of rainfall begins as the raindrop strikes the ground surface. Mihara (1951) has calculated that the kinetic energy of a raindrop amounts to 10^4 ergs for a drop of 2.5 mms. A drop of 2.5 mms accomplishes work equivalent to raising a body of 4.6 gmms by 1 cm. Mihara found that the modal size of rain-drops recorded in 25 rainfalls at Fukuoka and Tokyo in 1947 and 1948 was 0.4 mms, that drop size declined rapidly towards a size of 3 mms, but that most work was accomplished by the larger droplets. Best (1950) has discovered that there is a direct double-logarithmic relationship between raindrop size and rainfall intensity. Laws (1940) found that rainsplash erosion increased by 1200% as drop size was increased in a study using rainfall simulation equipment.

Mihara (1951) has made a photographic record of rainsplash which shows spray attaining heights of 0.7 m and distances of 0.15 m from the point of impact. These photographs also show that it is unusual for the spray to consist of water alone. Most particles have a soil particle nucleus. Ellison (1952) has demonstrated experimentally that a 5.9 mm droplet falling at 2.44 m/sec can move a 2-mm particle 0.2 m and a 4-mm particle to a distance of 0.1 m. Bisal (1960) has related splash detachment of soil particles to the 1.4 power of drop velocity. Free (1952) found that splash from sand varied with the 0.9 power, and from loam with the 1.5 power, of the rainfall intensity.

However, Farmer (1973) discovered that particle detachability was more strongly influenced by particle size, particle type, and the presence of a thin sheet of surface detention water than by land-slope or rainfall intensity. It was found that the peak detachability of particles from single grained granitic soils was 500μ while that for clay soils was 400μ and the detachability of the single-grained soils between 25-400μ was 1.5 times greater than that for clay soil aggregates (also: Woodburn and Kozachyn - 1956). Palmer (1964) has demonstrated that the presence of a thin surface layer of water increases raindrop impact forces though this effect inevitably declines with increasing depths of surface water. Rainfall duration is another factor which affects splash erosion. Ellison (1944, 1952) has demonstrated that maximum splash occurs shortly after surface wetting as a result of drop impact disrupting unstable surface aggregates. Subsequently, rainsplash detachment declines with time.

The nature of rain-splash erosion is also affected by slope angle, angle of incidence of rainfall and local wind velocity. Ellison (1944) has determined photographically that the soil moved down-slope by rainsplash may exceed that moved up-slope by a ratio of 3:1. Free (1952) has found that soil pans which face the oncoming rain can lose three times as much soil as those which face in the opposite direction. Disrud and Kraus (1971) have shown that soil detachment by wind driven rain is greater than that by similar rainfall intensities without wind. Wind reduces the buildup of liquid on soil clods prior to detachment. Consequently, it takes a smaller force to initiate detachment. It was shown that wind shear was up to half as effective as rainfall kinetic energy in initiating soil detachment.

De Ploey (1969, 1972) has measured splash erosion in several morphogenetic regions. He concludes that splash erosion is most effective and intense on the crest zones of convex slope and that its effect is to flatten them off. This effect is especially noticeable in semi-arid environments.

(b) Crusting

Only a third of the energy of raindrop impact is consumed in the generation of spray. The remainder contributes to the disruption and compaction of the soil surface. Soil crusting is a direct consequence of raindrop impact. Raindrop impact causes the compaction of the immediate surface (0.0-0.1 mm). It disrupts and disperses surface aggregates. In fact, a thin layer at the surface appears to slake away. The fine particles thus created are washed into the soil and fill the pores, cracks, and crevices near the surface. This process causes an increase in the bulk density of the surface layer of about 20% (Free - 1952; Lemos and Lutz - 1957).

Crust formation proceeds very rapidly during the first few minutes of rainfall (Epstein and Grant - 1967). The strength of the crust depends on the intensity of the rainfall. Low rainfall intensities produce crusts which are consistently weaker. However, the number of rainfalls is less important. Busch *et alia* (1973) have discovered that the strength of a crust may be influenced by a previous rainfall.

Soil crusts reduce infiltration. Tackett and Pearson (1965) have shown that the water permeability of soils underlying a soil crust may be five times greater than that of the crust itself.

Crusts will form on soils of almost any textural condition except those of a low silt/clay content. The strength of the soil crust is increased by a high soil clay content. Engelen (1973) has reported the following observations from clayey/badlands in South Dakota. Prior to rainfall, all the badland slopes were very dry and covered by a net-work of desiccation cracks. At the end of a storm which deposited 26 mm in 9 hours, however, only the uppermost 2-3 mm of the soil was saturated and plastic and only 50-100 mms below were at all moistened. The ground below remained perfectly dry. The swelling clays had acted as an efficient water seal (Warkentin *et al.* - 1957).

Surface crusts, therefore, are phenomena which result from raindrop compaction, soil structural change and the swelling and re-orientation of clay particles during surface wetting. Crusts consist of a seal, usually only 1-3 mms thick (Tackett and Pearson - 1965), and a thicker washed in layer which may achieve a thickness of 75 mms (Timms - 1971). The presence of a soil crust restricts soil removal but encourages increased runoff.

(c) Infiltration

Paradoxically, surface sealing by the development of soil crusts is not the only condition which may protect the ground surface from erosion by sheet flood. Loose granular soils such as those created by needle-ice activity (Soons - 1968) or those of freshly cultivated fields may suffer very little erosion under conditions of moderate to light rainfall. The reason for this is the high infiltration rates of such soils. The loose, open textured soil surface permits the rapid percolation of surface water and prevents the accumulation of water at the surface which initiates sheet flood. Of course, in a really heavy storm, even this enlarged infiltration capacity may be exceeded, and in this circumstance soil losses may be very heavy and there is a tendency for the soil to erode down to the base of its loosened layer.

Infiltration is, thus, an important parameter in the consideration of run-off erosion. Infiltration rates depend to a large degree on soil structure. Balogh and Matrai (1968) have proposed a simple division to separate soils of good and bad structure. Soils of good structure contain granules and aggregates of 1-10 mm size which are bound by lasting adhesive calcic colloids. These particles are capable of holding moisture over long periods and the rapidly drained interstices between contain sufficient air to support plant life and bacteria. Soils of poor structure, however, are composed of small particles and unstable aggregates which are dissipated easily and washed down to clog the soil interstices.

Infiltration also depends on the initial soil moisture content. The infiltration rates of damp soils tend to be far lower than for those which are dry. Subsequently, however, the differences between these two rates diminish. Balogh and Matrai (1968) have calculated the stable infiltration rates for several types of soil. The stable infiltration rate is a minimum rate of infiltration established after several hours of infiltration and, of course, is the same both for soils which were initially wet and those which were dry. This parameter is closely correlated with soil organic content and non-capillary porosity (Free <u>et</u>

al. - 1940). Balogh and Matrai's results are presented in Table 4.1.

TABLE 4.1

INFILTRATION RATES OF COMMON SOIL TYPES
(From Balogh and Matrai - 1968)

Soil Type	Infiltration (mm/hour)
Sand	+30
Sandy Loam	24-30
Loam	18-24
Clayey Loam	12-18
Clayey Soil	6-12
Heavy Clay	3- 6
Saline Clay	-3

The occurrence of surface runoff, therefore, depends on a number of other factors. Infiltration rates must be considered alongside initial soil moisture content, together with the nature of the soil and the soil surface, its ability to crust, slake and seal, along with the intensity and duration of incident rainfall. However, while the critical limits for the occurrence of surface runoff are very difficult to predict its incidence and nature are usually quite important and apparent.

(d) Surface Wash

When infiltration capacity is exceeded water begins to accumulate at the soil surface. Eventually, the coalescence of such water bodies may be sufficient to give rise to flow. The depth of this flow, in all probability, will be very thin, and its effective channel will be very wide (Chow - 1959). Vermes (1971) has compared this stream to a veil, but it is more properly termed "sheet-flow" or "sheet-flood."

The quantity of surface wash is governed by the width and depth of flow, the slope of the surface, and the resistance characteristics of the surface. Kowobari et alia (1972) have studied the effect of surface roughness elements on sheet flood. They discovered that an increase in the density or size of roughness elements tended to cause a proportionately greater increase in surface resistance to flow. Slope had very little effect on the flow resistance, however, it was found that an increase in discharge caused a decrease in surface resistance on relatively smooth surfaces, but an increase in resistance on surfaces with an increased roughness element.

The calculation of R (Reynolds Number: $R = \frac{VR}{v}$: where V is flow velocity [fps], R is hydraulic radius [ft], v is the kinematic viscosity of water [ft^2/sec]) from experimental sheet-flood data frequently gives very low values R = 0 to 130 (Kilinc and Richardson - 1973). The transition from laminar to turbulent flow for sheet-flood occurs at R = 300 to 800 (Chow - 1959). Now sheet-flood under rainfall conditions cannot strictly be described as laminar even though flow is influenced by viscosity and perturbations are damped. Raindrop impacts provide the flow with a continuous series of perturbations and thus the appearance of turbulence. This quasi-turbulent condition is termed agitated laminar flow and is usually super-critical, and may be equivalent to normal turbulence in its increased capacity to transport sediment (Kilinc and Richardson - 1973; Emmett - 1970).

Kilinc and Richardson (1973) recommend the following equation as a first approximation for soil losses resulting from overland flow generated by a single storm:

$$Qs = (1/e^{11.645})\ Re^{2.054}\ So^{1.460}$$

where: Qs is the total sediment discharge in pounds per foot width of slope surface, Re is the Reynolds Number, and So is the slope angle in percent. Kilinc and Richardson's (1973) evaluation of the efficacy of this equation is supported by a good high coefficient of determination ($R^2 = 0.94$) and good low standard error of the estimate (S.E.E. = 0.14) calculated from their rainfall-simulator studies.

The removal of surface particles is dependent upon the sheer strength of the surface materials, the resistance to flow offered by the surface, and the nature of flow across that surface. Since the resistance to flow is greatest on rough surfaces and surface roughness also helps generate turbulence which effects a greater surface transport of detached debris, there must inevitably be a tendency for sheet-flood erosion to create smooth surfaces which have least resistance to flow and thus the least potential for erosion.

Smith and Wischmeier (1957) have calculated that the kinetic energy of surface runoff is much the same as that of the total rainfall during runoff. Of course, this total must be much smaller than total rainfall kinetic energy. Rainfall has an important affect on sheet-flood. Rainsplash disrupts sheet-flood causing an increased turbulence and thus an increased potential for transport (cf: Shen and Li - 1973). Splash detachment probably supplies much of the load for sheet-flood transport since the total kinetic energy of rainfall is far higher than that for sheet-flood.

(e) Rilling and Gullying

It has been demonstrated that a large proportion of the transported soil detached during rain-storms moves by rolling or saltation on the floor of small flow channels (Foster and Meyer - 1972). Engelen (1973) has described the evolution of micro-rills on slopes which prior to rain were covered by a dense network of desiccation cracks and loose soil aggregates. The incidence of rain tended to close these cracks partly by swelling and partly by the downslope sliding of the surface aggregates. However, many of the cracks, especially those which had a downslope tendency were incorporated in a well integrated pattern of tiny, 2-5 mm wide by 2-5 mm deep, channels which formed a kind of trellis pattern with an internal spacing of 5-15 mms. These channels are very ephemeral and disappear once again as the slope is desiccated and the polygonal cracks and aggregates re-appear.

Desiccation cracks, however, are often much more obviously related to rilling. Observation of some recently created man-made clayey slopes on the banks of a tributary to the Pantano Wash in Tucson, Arizona and on a building site in the Palos Hills suburb of Chicago revealed polygonal trellis-like patterns of tiny, but relatively permanent, 10 mm wide, 4-7 mm deep, channels each floored by a milli-meter wide desiccation crack. The polygonal nature of this channel network is best revealed at the crest of these slopes. Further downslope, there is an increasing tendency for the portions of network which most nearly corresponded to the flowline slope to be deeper than the cross-slope portions. The foot of the Tucson slope is undergoing dissection by a series of deeper (15 mm) sub-parallel, regularly spaced finger rills. It is interesting to note that the distance separating these rills was similar to the average diameter of the desiccation crack polygons at the slope crest (70-80 mms) (Haigh - 1977: In preparation; and Figure 4.6).

As the coarseness of the surface materials increases so the permanence and stability of rill systems increases. Morisawa (1964), who studied the evolution of rill patterns on a newly upraised lake floor in Montana, discovered that the rate of development of rills was dependent on the initial slope of the land surface and the type of material. In general, it was found that the channels of streams on flat silty surfaces had wide, shallow, vertically walled, valleys, with intricate drainage nets and the tendency for tributaries to enter the trunk

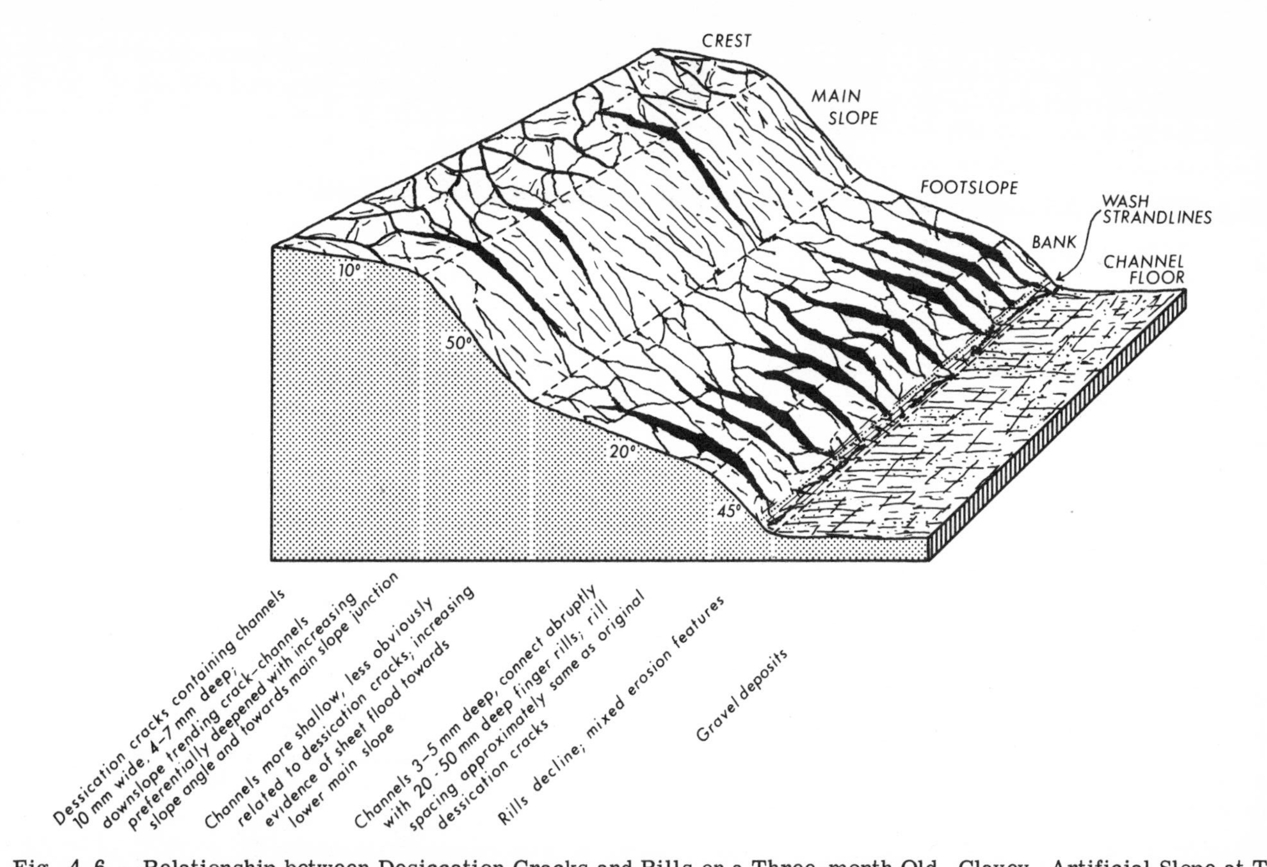

Fig. 4.6.--Relationship between Desiccation Cracks and Rills on a Three-month Old, Clayey, Artificial Slope at Tucson, Arizona.

at ungraded junctions. Again, headwater erosion proceeded by the widening of desiccation cracks. Such channels may be 10-100 mm deep by 50-5000 mms wide (Carson and Kirkby - 1973). Fine materials are, as has been already noted, more susceptible to freeze-thaw and moisture changes. There is a tendency for movements of the surface materials to obliterate these rill channels between flows. Rills which develop in sandy materials and on steeper slopes are straighter and deeper with V-shape profiles. They tend also to be more permanent.

Kilinc and Richardson (1973), in their rainfall-simulator studies of erosion by overland flow, found that rills occupied 18-45% of their surface area and caused 10-40% of the erosion. They devised two equations to predict the relative rill surface area and the volume of rill erosion:

$$\frac{\text{Area Rills}}{\text{Area Total}} = (1/e^{2.345})\ \text{Re}^{0.338}\ \text{So}^{0.144}$$

$$\frac{\text{Volume Rills}}{\text{Total Volume Erosion}} = (1/e^{2.927})\ \text{Re}^{2.054}\ \text{So}^{1.46}$$

where Re is the Reynolds Number and So is the slope angle in percent.

Schumm (1956a) has formalized the life history of the rill as an annual cycle. The rill is created during heavy summer rains and destroyed by subsequent winter frosts. Rills developing in successive seasons may, or may not, adopt the same channels. A rill in a cultivated field is a channel which may be seasonally destroyed by the plough. The essence of a rill is thus the fact that it is an ephemeral channel. A gully is a channel which persists from year to year.

The process by which a rill evolves into a gully has been described by Horton (1945). The initial development of rills on a slope usually appears as a system of essentially parallel and independent channels. It may be that some of these channels have better situations than the others, tend to redevelop in successive seasons, develop a small valley around themselves, and capture their more ephemeral and less active neighbors. A rill which develops in this fashion is termed a master rill.

An attempt has been made to classify gullies as continuous or discontinuous (Leopold and Miller - 1956). Continuous gullies begin, high on the slope, as a coalescence of numerous small rill channels and continue their courses right down to the valley floor. Discontinuous gullies begin as an abrupt headcut, which may be located at any point on the slope, and their channels may intersect the surface of the slope at any point (Heede - 1967, 1970, 1974; Blong - 1970). A feature of the channels of discontinuous gullies is that their slope is characteristically less than that of the interfluve. The depth of discontinuous gully incisions thus declines down-slope. Discontinuous gullies develop at a site of locally lowered resistance to erosion caused by trampling, grazing, fire, etc. An initial furrow is formed which is exaggerated by subsequent storms.

Discontinuous gullies may fuse to produce continuous gullies. In fact, most gully systems are hybrid between the two types. A continuous gully channel may thus contain internal headwalls and violent changes in depth while discontinuous gullies may be linked by small channels. A "type" continuous gully has no such discontinuities in its channel and its depth is either uniform throughout its length or more usually increases downslope from its headwaters to achieve a maximum at the junction of its interfluve main-slope with the lower slope element. The knickpoints of discontinuous gully systems are usually 0.2-0.5 m deep. The modal depth of a continuous gully is perhaps 0.6-1.5 meters. The maximum depth of a gully which increases in depth downslope may, however, exceed the range 2-5 meters. Its depth depends on the local slope configuration, its available relief (cf: Heede - 1970), and the mechanical properties of the eroding material (De Ploey - 1974).

Recent research has suggested that the planimetric growth of competitive gully systems conforms to the Horton/Woldenberg model of allometric growth (Faulkner - 1974; Woldenberg - 1966; Horton - 1945) and to the equation $y = ax^b$. Gully channels, however, tend to arise rapidly as the result of a particular disturbance or set of disturbances. Consequently, it is not uncommon to discover gully systems which are inactive or exist in a relatively steady state, as a result either of the cessation of the disturbing impetus or of morphological adaptation to the new hydrologic environment (White - 1963; Tuckfield - 1964, 1965; Blong - 1970; Nir and Klein - 1974).

(f) Pipes and Seepage Steps

The effects of concentrated flow are not limited solely to the ground surface. Sometimes, because of the development of large scale local differences in hydraulic pressure, underground waters become competent to carry small particles through the more permeable parts of a permeable formation. The development of these subterranean drainage tubes, which usually occurs in insoluble, usually fairly fine-grained, clastic sediments, is termed "piping" and the tubes themselves are pipes (Parker - 1963). Pipes are commonly associated with the presence of permeable and easily eroded materials and pipes are often translated into gully channels by the collapse of overlying soils. Sometimes, the partial collapse of the large scale pipes which may develop in semi-arid terrain gives rise to a pseudo-karst type topography. A common mode of pipe development is as subterranean headwaters for gullies eroding a densely turfed area. The pipes develop in biogenic or desiccation cracks occurring immediately beneath the turf root zone (Zaborski - 1972).

Jones (1968) describes the following sequence of events leading to the formation and destruction of a typical pipe system in the San Pedro Valley, Arizona. First, a steep-sided channel (arroyo) forms in erodible but cohesive sediments previously devoid of a distinct drainage channel (cf: Cooke and Reeves - 1976). This creates an increase in flow in fractures and subsurface channels which become enlarged beyond any possibility of closure by swelling. The pipe system grows out towards steeper parts of the catchment where runoff is more concentrated. This isolates parts of the original system, a process enhanced by changes in base-level emanating from the generating arroyo. Later collapse of pipes and sides leads to gradual exposure of a surface gully system. Technically, these gullies should appear identical to normal surficially produced gullies. However, if, as seems common, pipes occur on several levels, such gullies possess distinctive stepped cross and long profiles (Haigh - 1977: In preparation). Piping is a process which attains its greatest development in semi-arid environments (Fuller - 1922; Downes - 1946; Brown - 1962); however, a scaled down version of the piping process may develop under comparable initial stimulations in other climates (cf. Waun Hoscyn, Chapter 16).

There is often a close relationship between the evolution of rills, gullies, pipes and seepage steps. Seepage steps tend to be small crescentic erosional scarplets of perhaps 0.5-0.6 meters height. Their development is due to an interception of the ground surface and ground water table which may be associated with a local zone of low permeability (Hadley and Rolfe - 1955). The occurrence of this seepage causes the development of a free face because the local availability of water intensifies debris removal (Clayton and Tinker - 1971). Seepage steps may become exploited as the debouchment of subterranean pipes and/or gully headwalls. Erosion from stepped slopes is greater than that from smooth slopes (Brice - 1958).

(g) Soil Creep

Soil creep is a result of the disturbance of soil particles. Many of these disturbances take the form of cyclical changes in the found surface. Creep occurs because although vertical expansions of the ground surface may take place in a plane which is normal to the soil surface, gravity tends to lend subsequent contraction a vertical component and this results in a net downslope movement of the vibrating material. The simplest way to visualize this process is to consider the behavior of a soil particle raised by a column of needle-ice (Soons - 1968). The main constraint on the power of the vertical component is the packing of the soil particles. A soil particle may only adopt a path with a downslope component if that path is not blocked by another particle or some other body. Should that path be blocked by a growth of ground-ice or vegetative material, the resultant creep may be neutral or even negative – at least in the short term.

Quantitatively, the two greatest causes of soil vibration are the swelling and shrinking of the soils colloidal component on wetting and drying (Young - 1960) and the expansion of water on freezing which is normally accompanied by the electro-chemical migration of water to the freezing plane – usually initially located at the ground surface (Geiger - 1965) and the segregation of bodies of ground-ice or needle-ice (Everrett - 1963; Hoekstra - 1967). These effects may be exaggerated if different portions of the soil react differently to the different stimuli and may be assisted by subsidiary vibrations occasioned by simple expansion and contraction and the passing of heating and cooling waves into the soil (Geiger - 1965). The rhythm of these disturbances may be circadian or circa-annual. Schumm and Lusby (1963) have investigated the effect of the latter movements on the altitude of the ground surface as measured by stakes at Badger Wash, Colorado, a vegetated but semi-arid catchment. They discovered that stake exposure was least in spring after the soil had been loosened by winter frosts and most in the autumn after the soil had been compacted by summer rain-splash. The combined process of frost action, wetting/drying, and compaction effected creep in the upper 50 mms of soil. Clayton and Tinker (1971) have determined that creep is responsible for only about 1% of the retreat of the Badland slopes of South Dakota. However, it may be that on certain susceptible soil types (Schumm - 1956b) and on more vegetated slopes it may be the dominant process of slope alteration (Williams - 1973; Young - 1974). It is generally considered that soil creep leads to the development of slopes of rounded profile (Schumm - 1956b).

(h) Wind Erosion

The mechanics of wind erosion are similar to those which control the entrainment of soil particles by sheet-flood. Once again, particles moved by the wind tend to proceed by saltation or surface creep. Critical factors in the determination of the rate and nature of wind erosion are surface roughness, surface resistance, and effective wind shear. Murayama (1964) has shown that cultivated fields may develop a concave profile between wind breaks as a result of aeolian erosion. Pantastico and Ashaye (1965) have demonstrated that the presence of a surface stone layer, such as may also be developed by sheet-flood activity which tends to preferentially remove finer surficial particles, may reduce wind erosion or even cause the accumulation of wind transported materials. Wind erosion is not influenced by the soil moisture status of the surface except as it affects aggregate stability. An increased surface moisture content or the prolonged presence of snow cover (Anderson and Bisal - 1969) tends to preserve larger soil aggregates and reduce erosion. The presence of vegetation effectively eliminates surface wind shear and eliminates wind erosion (Geiger - 1965).

(VI) The Effect of Vegetation

Vegetation is a very important inhibitor of all slope processes. However, the presence of a discontinuous vegetation cover, perhaps the result of a local instability of the surface layers and the concentration of erosive pressures, may act to exaggerate the morphological impact of the particular processes which prevent revegetation.

Baver (1956) discusses the effects of vegetation under five headings. These are: (1) the interception of rainfall by the vegetative canopy, (2) the decreasing of the velocity and cutting action of surface run-off, (3) the root effects in increasing granulation and porosity, (4) biological activities associated with root growth and their influence on soil porosity, and (5) transpiration of water leading to the subsequent drying out of the soil.

Several further factors may be added not least among which are the creation of the humus layer and its effect on erosion and the stabilizing effects of root systems. Riedl (1970) claims that within each cubic centimeter of forest soil there may be 60 centimeters of root. The interpenetration of branching root hairs effectively re-inforces and binds the soil particles of the upper soil horizons. In forests, the stabilizing influence of roots may affect considerable depths: -2 m under beech, 0.6 m under spruce, 0.3-0.5 m under bushes. Smaller plant species tend to affect progressively smaller depths. These influences tend to restrict both failure and creep processes.

Again, the thicker and more dense the vegetation cover, the better its capacity for rainfall interception. Thus, while a well developed spruce canopy may intercept 33% of the annual precipitation, beech will only intercept 15-20%. However, such a canopy may intercept 35-43% of a single shower (Reidl - 1970). Wollny (Baver - 1938) has demonstrated that crop plants may intercept 12-55% of incident precipitation depending on the type of plant and number of plants per unit area. The importance of rainfall interception is not that it prevents the water from reaching the ground. It is that it reduces the kinetic energy of the fall by reducing the incident rainfalls terminal velocity. The water which drips from leaves or flows from stems is incapable of reducing infiltration capacity by compaction or of causing much splash transport. Further, the fact that intercepted rainfall is delayed from reaching the ground means that the supply of water to the ground surface occurs over a longer period. This means that there is less opportunity for accumulations of water to build up at the surface because there is a longer time available for percolation and infiltration.

The presence of vegetation inhibits surface runoff. Shoot systems, leaf litter and protruding roots all contribute to the hydraulic roughness of the soil surface (vide: Li and Shen - 1973). The root tissues absorb moisture to supply liquid for transpiration and also hold moisture on their surfaces. The numerous micro-organisms of each rhizosphere act in a similar fashion as do the free-living but associated soil organisms. The passage of roots and the migrations of soil fauna create channels in the soil which greatly facilitate infiltration. The net result of this activity is that while the runoff from a bare soil surface may be 80% (Meyer et al. - 1971), that from meadow grassland may be as little as 7%, of the incident rainfall (Wischmeier - 1970). Dragoun (1969) discovered that the provision of a vegetative cover of seeded perennial grasses to a water shed in the Great Plains area caused a 90% reduction in the amount of runoff inside two years. Observations made during actual rain-storms on unvegetated slopes show that while the runoff from clayey badland slopes may approach 80-90% of the incident rainfall (cf: Engelen - 1973), that from a thinly vegetated, semi-arid watershed may be more like 50-60% (Hadley and Lusby - 1967), and that from temperate agricultural land may be 20-30% from a field planted with root crops, and less than 0.1% for meadows or fields with well grown grain-crops (Gil and Slupik - 1972). Naturally, these results are only crude estimates since the amount of recorded runoff depends also on the initial soil moisture content and infiltration capacity as well as the intensity and duration of the storm. Nevertheless, such reductions in surface runoff have an important effect on the amount of soil removal. Reidl (1970) reports that to plant an unprotected soil surface with alfalfa reduces soil loss by 81.3%. Gil and Slupik

(1972) found the following variations in the ground retreat recorded from slopes planted with different crops during a single storm in 1969. The retreat recorded from a potato field was 4.0 mm, that from a meadow 2 x 10^{-6} mm, that from a field of rye 6 x 10^{-7} mm. Results from a year's observation demonstrate that the annual retreat of areas planted to root crops in the Bystrzanka catchment was 2.97 mms p.a., of that planted to cereals and meadow, 0.0043 mms and 0.0028 mms p.a. respectively, and of that under forest, 0.000007 mms p.a. (Gil - 1974).

Dietrich and Meiman (1974) have assessed the importance of plant transpiration in reducing the water potentially available for runoff in a study of patch-cutting in the Lodgepole Pine forests of the Colorado Front Range. They discovered that the average increase of water potentially available for runoff was 118 mms. Seventy-nine percent was attributable to reduced evapo-transpiration resulting from tree removal.

Kilinc and Richardson (1973) produced the two diagrams displayed as Figure 4.7 to summarize their laboratory investigations of the effect of vegetation. The diagram on the left shows that increasing rainfall intensity reduces the erosion-retardation effect of vegetal cover of winter wheat by decreasing the relative effect of impact energy, flow resistance and flow velocity. The diagram on the right shows averaged sediment concentration from a vegetated and unvegetated surface at 40% (22°) slope with varying intensities of rainfall.

The nature of the vegetation and land use also affects the disposition of retreat across the slope profile. The results of Gerlach's studies in Poland (Table 4.2: Gerlach - 1966), illustrate how the conversion of forest to pasture or arable land not only accelerates the rate of ground retreat but initiates a different pattern of slope evolution. Gerlach's results show that the conversion of a forested slope suffering uniform ground loss to pasture causes a massive acceleration of retreat on the lower slopes but a much smaller acceleration on the upper slopes. A retreat differential is established such that lower slopes retreat 42.6 times as fast as the upper slopes. This situation implies that the slope will tend to become steeper as a result of its changed land use. The conversion of forest to arable land causes an even greater acceleration of erosion but, here, that acceleration is greater on the upper slopes which retreat 1.9 times faster than the lower slopes which implies a tendency for slope decline (Gerlach - 1966).

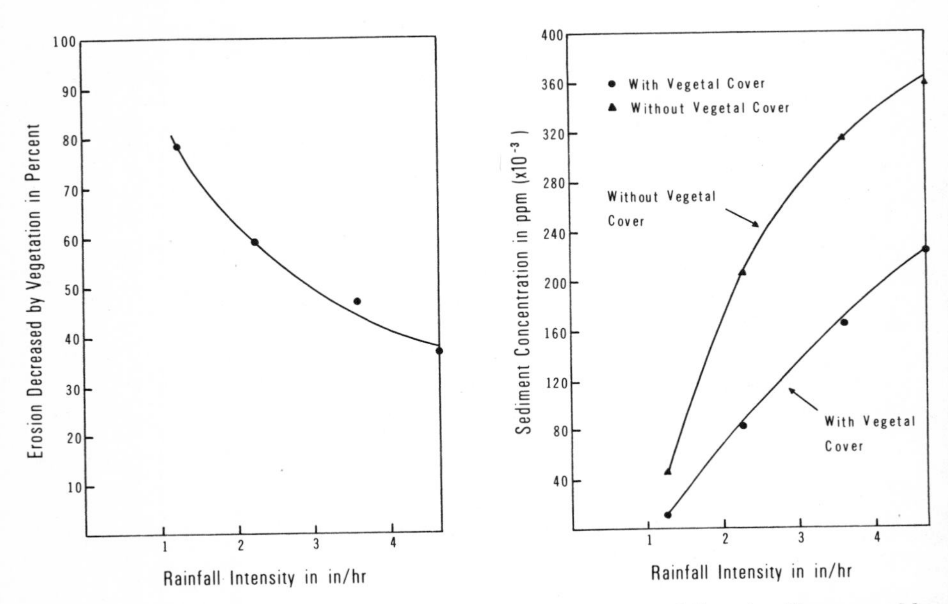

Fig. 4.7.--The Relationship between Percentage of Erosion Decreased by Vegetation and Rainfall Intensity on a 40-percent Slope Covered with 3-4 Inch High Winter Wheat and the Relationship between Sediment Concentration and Rainfall Intensity on a Vegetated and Unvegetated 40-percent Slope.

TABLE 4.2

SOIL EROSION VALUES AT THE EXPERIMENTAL STATION
OF THE POLISH ACADEMY OF SCIENCES
IN THE POLISH CARPATHIANS
(T. Gerlach - 1966)

Vegetation Type	Absolute Erosion Mm/year		Relative Erosion	
	Upper Slope Section	Lower Slope Section	Upper Slope Section	Lower Slope Section
Forest	0.00003	0.00003	1	1
Pastureland, meadow	0.00075	0.032	25	1,066
Field	0.9	0.47	30,000	15,666

(VII) The Effect of Trampling

Trampling affects slope evolution in two ways: compression and disruption. Selby (1972, 1973) has demonstrated that land which has been subjected to trampling as by the pasturing of grazing animals has a markedly lower infiltration capacity than natural grassland. His studies on North Island, New Zealand show that mean runoff from pasture plots was 4.6% of incident precipitation but that that from ungrazed grass and scrub was 0.9% and 0.7% respectively (Selby and Hoskins - 1973).

The subject of the origins of terracettes is a vexed question. Several authorities (Tivy - 1962; Shchukin - 1973) suggest that these features are primary wave-like forms which are due to gravitational sliding and the solifluction of sub-soil layers. Latterly, the flatter portions of these features are selected as preferred routes for movement and subsequently attain the status of true terracettes. Two points are in need of further consideration. First, terracettes are not sheep tracks. These are separate and quite distinct features. Second, it is most unusual to record examples of terracette development from slopes which are not subjected to a considerable trampling. Terracettes are a distinctive micro-relief form. Characteristically, their treads possess a width of 0.2-0.4 meters and their, usually vertical, riser a similar height. It is possible that terracette dimensions are affected by local slope angle but this contention has not yet been investigated.

Foot-paths and tracks are another characteristic feature of trampled areas. Trampling gives rise to distinctive plant communities and heavy trampling entirely destroys the plant cover. The devegetated sections of foot-paths and related features are frequently areas of concentrated erosion and, often, gully initiation (Bates - 1935; White - 1963; Evans - 1973; Streeter - 1975).

Trampling may also cause the disruption of sharp turf edges and thin surface soil crusts through the localized pressures of sharp hooves. Such disruptions of the surface inevitably lead to an increased erosion of the affected areas

Sometimes terracettes and sharp turf edges are exploited for shelters by animals, especially sheep. Sheep burrowing, caused by the rubbing of the animal as it huddles in its shelter, can be an important agency in slope erosion especially in areas which support a dense sheep population (Thomas - 1965, and Photograph P.4.1: Sheep burrows at Milfraen). A similar, if smaller, impact is made by other species, notably the rabbit. Bare scars may develop near the entrances to warrens and the underground systems may be important in the transmission of underground waters.

(VIII) The Effect of Aspect

It has long been recognized that slope aspect influences the rate and mode of slope development. Wollny in 1887 (Baver - 1938) discovered that runoff was greater from slopes in the following aspect classes according to the following series: north-west-east-south. The fundamental reason for this situation is differential insolation. Geiger (1965) has summarized the literature which demonstrates that in the northern hemisphere slopes which face north-north-east tend to receive less insolation than slopes which face south-south-west. The degree of this insolation differential is a function of the angle of incidence of the sun's rays at the ground surface, a function which varies with latitude, slope angle, and of course slope orientation. This differential heating of opposed slopes effects a differentiation of slope processes through the agencies of promoting differences in soil moisture content, vegetation cover and type, and freeze-thaw activity. The differential activities of slope processes affected by these variables leads to the differential development of soil (Bridges - 1961; Stepanov - 1967) and the differential development of slopes, both directly and via the process of channel migration. These phenomena are most frequently discussed in the context of the widespread phenomena of valley asymmetry (Karrasch - 1970, 1972; Haigh - 1973; Kennedy - 1973).

There have been several direct measurement studies which have detected the influence of slope aspect as it affects direct records of ground retreat. Tinker (1970; Clayton and Tinker - 1971) has reported that the average lowering of slopes underlain by the Tongue River formation in the North Dakota bad-

P. 4. 1. --Sheep Burrowing on the Milfraen Spoil Tip Complex

lands was 3.6 mms p.a. on slopes which faced southwest but only 2.8 mms p.a. on slopes which face northeast. Crabtree (1971) has described the asymmetrical infilling of the ditch section of the experimental earthwork at Overton Down. It was found that overall the south-west facing slope retreated and became shallower in angle more rapidly than the north-east facing slope. The south-west facing talus slope also contained finer debris. However, it was noted that there was a marked variation in the balance of debris production from year to year and it was far from unusual for the annual addition of debris to the north-east facing slope to be finer than that deposited on the slope which faced southwest. Crabtree (1971) suggests that the main agency responsible for the differences in the two profiles was freeze-thaw. The importance of freeze-thaw activity, and needle-ice formation, was also emphasized by Schumm (1956a) in his discussion of the rill cycle as being an important factor in the differentiation of erosion by concentrated wash upon slope of opposed aspect.

(IX) Conclusions

Current studies in slope research tend to tackle the problems of defining slope evolution from one of about four stand-points:

(i) The traditional approach, deductive and mathematical modellings based, uniformly, on an insufficient appreciation of the complexity of the field situation is represented by a large and sterile literature. Passable slope profiles may be generated by any number of equations based on an infinitely variable suite of initial assumptions. There is, of course, always a chance that one of these equifinite solutions is correct;

(ii) The morphometric approach, based on an attempt to describe and classify the geometric properties of slope profiles and chart their evolution in terms of subsequent geometric alterations, is rather more fruitful. Such a taxonomic approach to slope research is unlikely to provide any final solutions, but it would seem to be the most profitable and hopeful direction in current research;

(iii) The process approach, represented by a vast and heterogeneous literature including laboratory studies and field investigation. The aim is to analyze the role of the individual processes and control components governing erosion and hence slope evolution, then to study their inter-actions and relative importances. Some of the work produced in these investigations is of considerable significance. However, slope evolution is a divinely complicated process and many of its components are very imperfectly understood. Synthesis, thus, is a very long way ahead;

(iv) The empirical soil loss approach, based on the direct measurement of erosion and slope evolution, often in laboratory conditions, but, happily, sometimes in the field. It has proved the most satisfactory, practical and valuable line in current research. This approach, pioneered by the American Agricultural Engineers, has recently begun to produce studies which can not only assess the quantity of erosion experienced by particular slopes in particular conditions but also the distribution of such erosion. Already, attempts are being made to use these data to design "erosion-free" slopes, and one feels that the move from this stage to a gradual integration of the results of the process studies may not be too long delayed.

This study aims to continue the works begun by the empirical soil-loss researchers. Slope evolution is examined in terms of the operation and effects of a set of fundamental environmental controls. These controls include the effects of aspect, vegetation, biogenic disturbance, initial slope morphology, and the efficacy of basal removal. The action of these controls are, whereever possible, interpreted in the light of known soil properties and the operations of slope processes. Slope evolution itself is not described in terms of past theoretical models but expressed purely in terms of the geometric alteration of the relationships between individual slope units.

CHAPTER 5

CLIMATE RECORDS FROM THE BLAENAVON AREA

The preceding chapter stressed the importance of climatic controls on th nature and rate of operation of erosion processes. It is therefore of some importance to place these studies of erosion and slope development in some sort of climatic context. Unfortunately, the Blaenavon area possesses very few climatic records. This is a problem which has confronted several workers and a sort of solution has been suggested by Thomas and Kegie (1953). These workers employed climatic records from the recording station at Tredegar, in the next valley, to eke out the sparse information from the Upper Afon Lwyd.

At the present day, the climatic instrumentation established at Blaenavon consists of just three rain gauges. Two of these, at Llanelly Hill and Cwmavo have been operative throughout the experimental period, June 1972-June 1974. The third at Bunkers Hill was not established until April 1973. The Tredegar station which was operative at the start of the experiment was closed in March 1974.

Two of the Blaenavon rain gauges are operated by the Welsh National Water Development Authority. These gauges are located at Llanelly Hill, Wau afon (O. S. Grid Ref: 225106) at a height of 432 meters, and at Bunkers Hill near Blaenavon Town (O. S. Grid Ref: 252101) at a height of 457 meters. Rain fall records are gathered at these gauges at intervals which range from about five days to about a month. The rainfall recorded during the first year of study was 1448 mms at Llanelly Hill; during the second year of study it was 1204 mn at Llanelly Hill and 1086 at Bunkers Hill. The monthly totals of rainfall recorded at Llanelly Hill are graphed as Figure 5.1.

The rain gauge at Cwmavon is located on the floor of the Afon Lwyd valle to the south of Blaenavon at a height of 245 meters (O. S. Grid Ref: 269072). The Afon Lwyd valley at this point is very deeply incised and rather narrow sc there is the possibility that the data from this gauge are influenced by a rainshadow from the valley sides. Rainfall totals from Cwmavon are recorded on daily basis. The annual record from the first year of this study was 1310 mms and that from the second was 1276 mms. The monthly totals of rainfall re-

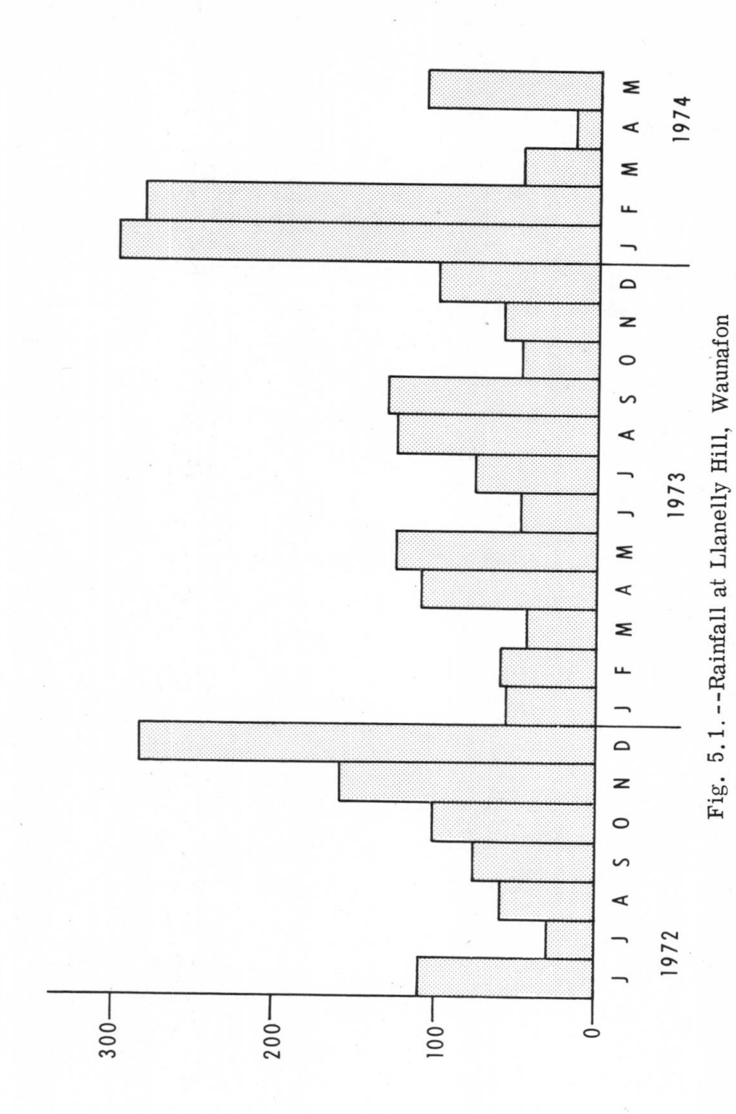

Fig. 5.1.--Rainfall at Llanelly Hill, Waunafon

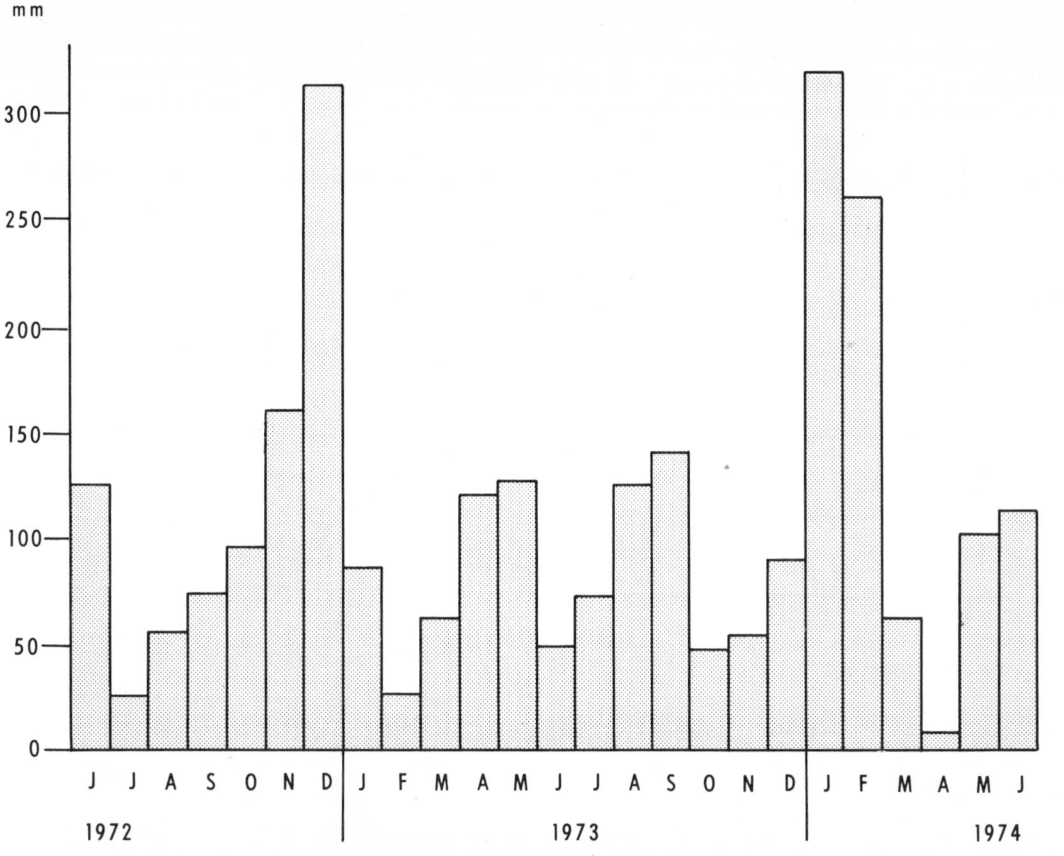

Fig. 5.2.--Rainfall at Cwmavon

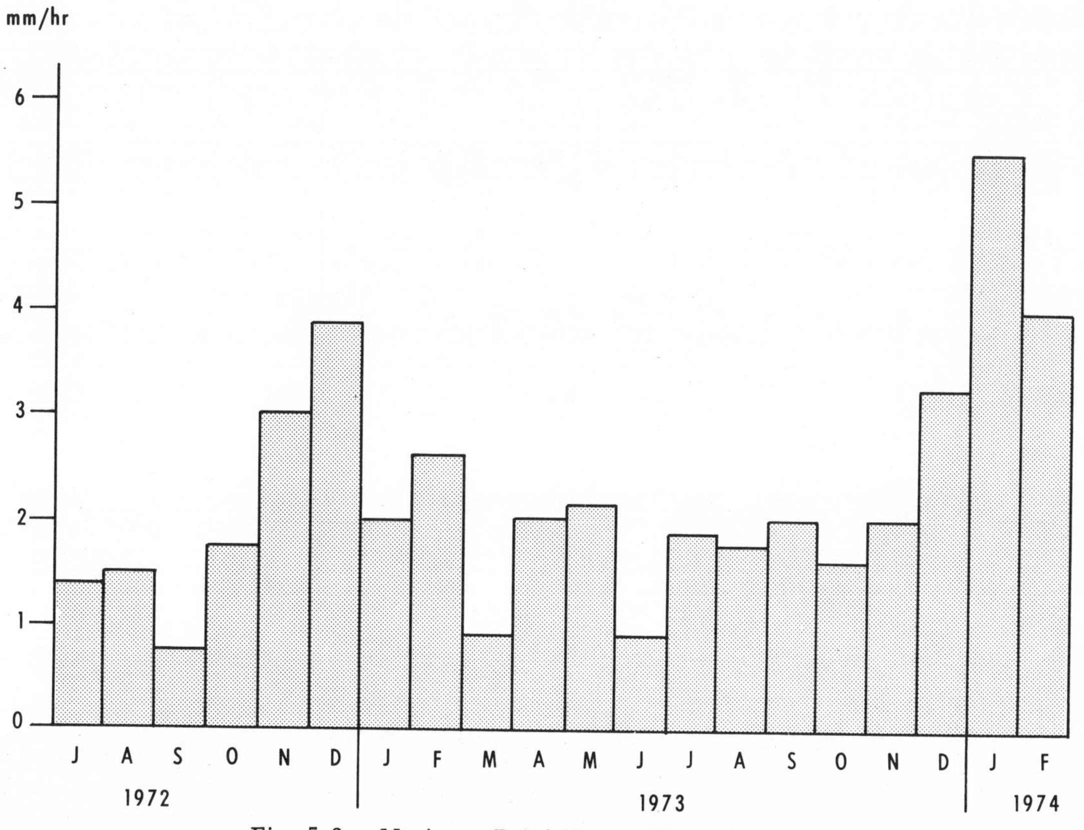

Fig. 5.3. --Maximum Rainfall Intensities at Tredegar

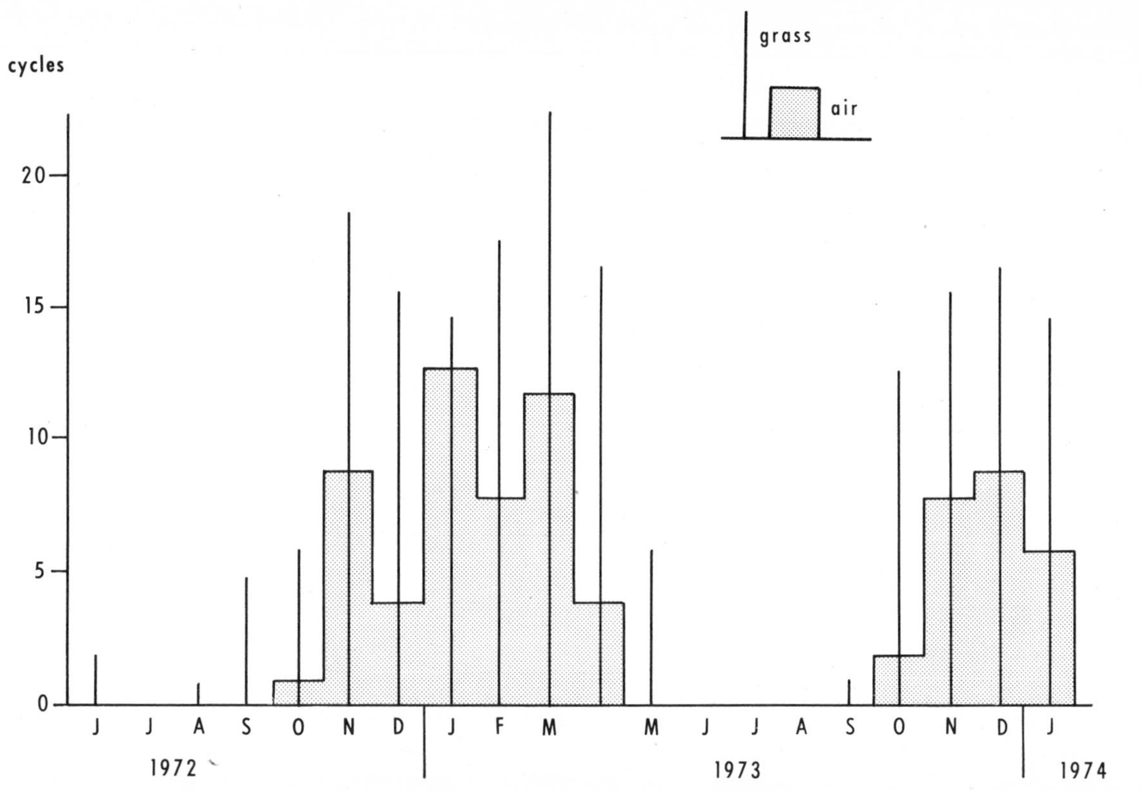

Fig. 5.4.--Freeze/thaw Cycles at Tredegar

corded at Cwmavon are graphed as Figure 5.2.

The results from Blaenavon's raingauges demonstrate that the area received its maximum rainfall in the winter months. Figure 5.3 is a graph of the monthly maximum recorded rainfall intensities recorded at Tredegar. These data suggest that highest rainfall intensities also occur in the winter months and that rainfalls of higher intensity were more common in the second year of the experiment.

There are no records of ground surface or soil temperature from the Blaenavon area. Figure 5.4 is a graph depicting the frequency of freeze-thaw oscillations at Tredegar. It is shown that the winter of the first year of the experiments was a period of much greater freeze-thaw activity than was the second year of the experiment.

CHAPTER 6

EXPERIMENT DESIGN

This study is concerned with the "wild-becoming" (Devdarijani - 1954) of the directly-created, constructional class, landforms of the Blaenavon area. The particular province for investigation is slope development on those artificial landforms which were created during the course of coal extraction, for this has been the dominant incentive for anthropogenic landform generation in the region.

The artificial landforms of this sort found in the Blaenavon area are of two main types and within each type may be recognized at least two sub-types. The main division separates those landforms which have been created by the dumping of deep-mined coal spoil from those which have been created by the infilling of an open-cast pit.

Deep-mined spoils tend to consist of black coal shales and may contain a high coal content. They are dumped as discrete landforms which may be visually obtrusive but cover a relatively small area. There are several divisions which may be suggested for the separation of different classes of these landforms. That employed here is the distinction between loose tipped spoils and those which have been created by tipping and compaction.

The spoil employed to infill an exhausted opencast pit is less homogeneous than that produced by deep-mine operations. It contains little coal and fewer coal shales. Its composition is mainly determined by the nature of the strata which overlaid the productive seam. At Blaenavon, much of this overburden consists of sandstones and clays. The landforms created from this spoil tend to occupy far larger areas than those due to deep-mine spoil. Further, the final surface of these features is not produced purely by tipping. It is customary for the surface of infilled opencast sites to be subjected to some kind of grading and some attempt to restore pre-existing water-courses. The subdivision of these land-features is not based on a consideration of their creation nor on differences in the nature of the overburden infill. The division suggested by this study is based on the consideration of environmental independence. Thus, while some of the former opencast sites at Blaenavon occupy former crest segments and are essentially independent of their surrounds, one opencast site occupies a mid-slope site and functionally might be considered as part of a larger slope system.

The basic over-plan of this study ensures that the processes of slope evolution are monitored at, at least, one location on representatives of each of these four classes of landform. Slope evolution is studied by means of instrumented slope profiles. Each profile is described by reference to its site and by a code letter which usually indicates its role in one of the secondary experiments. Twelve instrumented slope profiles are established and, in fact, at least two are assigned to each class of landform. The disposition of these experiments in the experimental area is displayed as Figure 6.1. The place of these experiments in the primary experiment plan is illustrated by Figure 6.2.

Beyond these basic divisions, this study recognizes a further sub-division of the class of tipped and compacted landforms created from deep-mined spoil. This distinction separates the class of low, and relatively old, fan-ridge spoil tips from their larger and more modern counterparts, the plateau spoil tips. It is argued that for many purposes the evolution of the smaller and older tips may be considered in the light of the studies of the younger plateau tips. However, an attempt has been made to support this assertion and to analyze the particular features of this sub-class by means of detailed morphometric survey.

The plan of this experiment also contains a provision for the study of one important internal control of slope evolution on the various classes of artificial landform. This is the presence or absence of a vegetation cover.

The secondary concern of this study is the study of the process of natural hill-slope evolution. The anthropogenic landforms considered by these analyses may be regarded as simple, idealized hill-slopes. The internal variations in these slopes' internal erosional environments is used to frame a number of experiments which have bearing on the wider problems of natural slope evolution.

The primary division which is recognized is between slopes which possess a basal gully control and slopes which do not. Broadly, it may be stated that all the slopes developing on deep mine spoil mounds, with the partial exception of Milfraen E., possess no basal gully control, while all the slopes on open-pit infill possess a basal gully control.

The second division, again, separates those slopes which support a vegetation cover from those which, in general terms, do not. Further subdivisions examine the effects of slope aspect, and recognize the importance of the variations in the land-forming materials in limiting the comparability of instru-

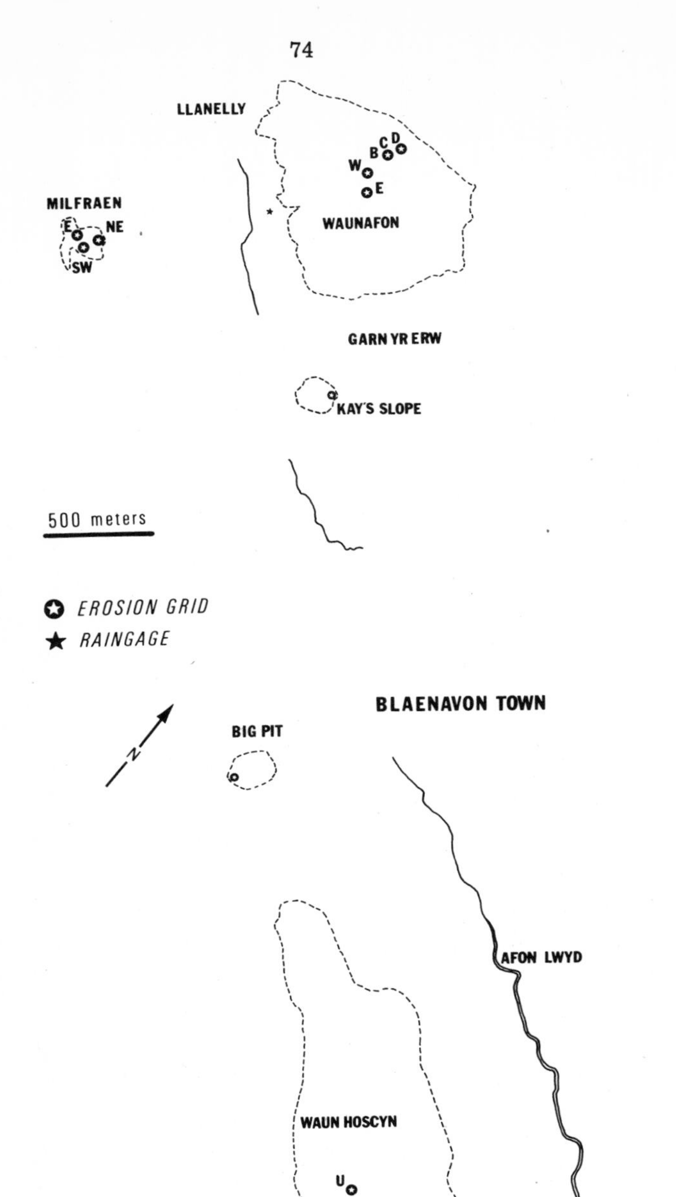

Fig. 6.1.--Relative Location of Erosion Grids

Landform Classification		CONSTRUCTIONAL LANDFORMS (created during coal extraction)			
		Deep-Mine Spoil		Open-Pit Infill	
		loose tipped	tipped and compacted	environment independent	environment dependent
Experimental Slopes	unvegetated	Big Pit Kay's Slope	Milfraen E.	Waunafon B Waunafon C Waunafon D	Hoscyn V. (dense seeded turf)
	vegetated	None	Milfraen N. E. Milfraen S. W. (supporting studies of fan-ridge Tip Morphology)	Waunafon E. Waunafon W.	Hoscyn U. (sparsely vegetated)

Fig. 6.2.--The Examination of Artificial Constructional Landforms

Slope Classification		SIMPLE ARTIFICIAL SLOPES			
		Basal Gully Control		No Basal Gully Control	
		vegetated	unvegetated	vegetated	unvegetated
Aspect Classes.	neutral	Waunafon E. Waunafon W. Hoscyn V.	Milfraen E. Waunafon C. Waunafon D. Hoscyn U.		Milfraen E.
	warm			Milfraen S. W.	(Big Pit Tip)
	cold			Milfraen N. E.	(Kay's Slope Tip)

Fig. 6.3.--The Examination of Controls of (Interfluve) Slope Evolution

mented slopes. The broad plan for this experiment is displayed as Figure 6.3.

The third concern of this study is the assessment of the importance of secondary slope gullies by the comparison of parallel slope and gully profiles. Four gully channels are examined here, together with their banks and the parallel interfluve slopes. Three of the gullies examined are evolving on unvegetated profiles, one profile is vegetated. All the gullies are situated on slopes of neutral aspect. However, the three gullies incised on unvegetated slopes are located on landforms of different types. Thus, one instrumented gully is located on the unvegetated slopes of the tipped and compacted plateau tip at Milfraen, one is on the unvegetated slopes of the environment-independent Waunafon infilled opencast site, the third is located on the spoil banks of the Waun Hoscyn Opencast. These spoil banks are located on a valley bench of the Afon Lwyd valley and it is the natural slope which provides the crest segment for the opencast spoil. The fourth instrumented gully is also located at Waun Hoscyn on a seeded grass sward perhaps 300 meters distant from the instrumented gully situated on the relatively unvegetated slope.

CHAPTER 7

TECHNIQUES I: EROSION PINS*

The erosion pin is one of the simplest and most effective methods for monitoring the minute changes in the altitude of the ground surface which are due to erosion and deposition. Essentially, an erosion pin is a benchmark. The head of the pin is counted a fixed reference and changes in its elevation are interpreted as changes in the height of the surrounding ground surface. A reduction in erosion pin exposure is termed "ground advance," and an increase, "ground retreat." Many research workers would regard these phrases as being synonymous with deposition and erosion, but this is not necessarily the case. Ground advance and ground retreat may occur independently of any erosion or deposition as a result of cyclical expansions and contractions of the ground surface due to heating and cooling, wetting and drying, freezing and thawing, and/or the hydration of clay minerals. Secular changes may also occur as a result of soil creep or compaction.

The questions which arise from the use of erosion pins may be defined thus:

(i) how fixed is an erosion pin?

(ii) how accurately can one assess its height above the surrounding ground surface?

(iii) how does one interpret the changes in elevation of that ground surface (i.e.: how much erosion/deposition, how much cyclical changes of the ground surface, how much soil creep/compaction, how much operator disturbance)?

(iv) how much does the presence of the erosion pin affect the course of ground advance and ground retreat?

Research workers have never entirely agreed on the answer to these questions

---

*N.B.--A modified version of Chapter 7 will be published as a contribution to the British Geomorphological Research Group's Technical Bulletin 18, 1977, Shorter Technical Methods, under the title: "The Use of Erosion Pins in the Study of Slope Evolution."

and the result of this lack of agreement has been a proliferation of different species of erosion pin experiment. Several factors seem critical to this variation and these include:

(i) the length and composition of the erosion pin;
(ii) the use of a washer;
(iii) the initial exposure allowed to the erosion pin;
(iv) the disposition of the erosion pins on the hillslope;
(v) the time interval between data records;
(vi) the method of recording.

(i) The Length and Composition of the Erosion Pin

The original erosion pins were wooden. Both Colbert (1956) and Schumm (1956a) employed wooden stakes for their studies of the erosion of badlands in North America. Wooden stakes were also employed by some early studies of snow ablation. More recently, Ranwell (1964) has confirmed the use of bamboo canes for the measurement of sediment accumulation on coastal Spartina marshlands. However, most workers regard wooden pins as too fragile (Colbert - 1956) and liable to rot, and, consequently, most studies advocate the use of metal pins. Metal pins tend to rust and are more liable to frost heave. Bridges (1969) claims that rusting tends to bind erosion pins into the soil and grants an increased resistance to disturbance. However, Haigh (1974a, b) has reported that major problems may result from the use of rustable materials in longer term studies. Washers seem to provide very suitable sites for the growth of ice crystals. Should the washer and rod become rust-bound, both become very liable to frost lifting. The adoption of a very loose fitting washer may avoid such difficulties (Kirkby and Kirkby - 1974) but further problems result from the rust-corrosion of erosion pin and washer which alters the height of the former and the width of the latter. Clayton and Tinker (1972) have recommended the use of non-rustable materials. Evans (1967) and Imeson (1971, 1974) employed brass erosion pins.

(ii) The Use of a Washer

There are a number of good reasons why many research workers advocate the use of washers in conjunction with erosion pins. They aid the relocation of the erosion pins and average out the unevenness of the soil around the erosion pin and thus allow a much greater standardization of recorded data.

Further, it is suggested, the use of a free standing washer allows one to record both maximum and net erosion. Erosion undermines the washer which slips down the erosion pin's shaft. Subsequent deposition occurs on top of the washer. The difference between the height of the soil surface and the buried washer is then taken to represent the difference between net and maximum erosion. Unfortunately, washers are not well designed as sediment traps, nor do they easily slip down the shaft of an erosion pin. Results based on this assumption, particularly when they are gathered from observations of a vegetated or stoney slope, should be treated with caution.

Clayton and Tinker (1972) claim that the use of a washer helps the observer to distinguish between real erosion and retreat due to cyclical contractions of the sediment. They note that when erosion occurs, pedestals of sediment are formed beneath each washer. Similarly, when deposition occurs there are traces of sediment on the washers. Where no such evidence exists changes in exposure are attributed to cyclical forces.

However, the presence of a permanent washer does influence soil erosion in its immediate vicinity. Permanent washers protect the ground beneath from rainsplash. Consequently, after heavy rains washers may be found perched some millimeters above the level of the surrounding soil. The existence of these pedestals further testifies to the fact that permanent washers interfere with the pattern of sheet-flood in their vicinity. If the washers are to be used as a standardized ground surface, these pedestals have to be ground out prior to recording. This risks disturbance of the erosion pin and causes the ground beneath the washer to become unusually compacted.

Schumm (1967), therefore, has proposed the use of a removable washer. This would be lowered over the erosion pin immediately prior to recording and removed immediately afterwards. Its use would eliminate the difficulties of measuring directly onto the soil surface and allow one to gather a record from an area beyond the immediate sphere of influence of the erosion pin. Unfortunately, readings gathered by such a method are inevitably more variable than those taken from permanent washers and are also more liable to influences arising from variations in the compressibility of the soil.

(iii) The Initial Exposure Allowed to the Erosion Pin

There are two problems connected with the choice of an initial exposure for the erosion pin: burial and accuracy. Many studies, especially those associated with the Vigil Network, recommend that, in the first instance, erosion pins should be established flush with the ground surface. Some studies recommend that the erosion pin be reset flush to the ground surface after each record (e.g.: Emmett - 1964). This is fine in a situation where one can guarantee no deposition. However, there are very few sites on even the most rapidly eroding slope where deposition cannot occur (Pinczes - 1971; Marosi - 1972; Haigh - 1974c). A buried erosion pin is difficult to locate and to record. However, it is not a good idea to allow the erosion pin to become too exposed since this increases the scope for error. It is easier to identify an increase of 1 mm in a total of 15 mms than it is to detect a change of 1 mm in a total of 100 mms. Further, increased exposure can make the pins top heavy and more liable to disturbance by tilting.

(iv) The Disposition of the Erosion Pins upon the Hillslope

Erosion pins may be established as a transect line down a slope or along an interfluve, or in a grid pattern across the slope, or as clusters at sites of special interest. Obviously, the precise format adopted by the individual research worker will depend on the nature of the experiment to be undertaken. Nevertheless, it is generally considered desirable that the record reported from any individual site is the mean result from several erosion pins. This precaution is intended to reduce random errors incurred during recording and to compensate for the extreme individuality of the erosion record from certain individual pins.

(v) The Time Interval between Data Records

The more recordings made at a given site, the more accurate the ground retreat record, and the more accurate the final estimation of erosion as random errors and irrelevant cyclical fluctuations tend to average themselves out. Conversely, the more recordings made, the greater the opportunity for the operator's own activities to influence the erosion record. The time intervals employed by the various research workers vary from 7 days (Bridges - 1969; Bridges and Harding - 1971) to over a year (Schumm - 1956b). Clayton and Tinker (1972) recommend that recording should be at least semi-annual and after every major period of runoff.

(vi) The Method of Recording

The normal method of recording is by the direct and usually repeated measurement of the distance separating the head of the erosion pin from the ground surface or washer. Nevertheless, several variations have been suggested. Disecker and Richardson (1961, 1962) located their erosion pins in two parallel rows, 1.5 m apart. Recording was effected by the use of a specially designed rod and bar device which was used to make measurements at 0.3 m intervals across the tops of all squares and diagonals. To date, this technique has only been applied to studies of scour in drainage ditches, but the approach would seem entirely appropriate to studies of slope and gully development. Campbell (1970 et seq.) has continued this idea by adapting a "Schefferville Frost-heave Bedstead" (Matthews - 1962) for the measurement of erosion. A light-weight aluminum frame was welded into a lattice configuration and 25 holes at 0.25 m intervals machined in a 1 m grid. Sliding aluminum rods were then fitted into these holes. For recording, this apparatus is mounted on four modified stakes, the sliding rods lowered to the surface and measurements read from them to the nearest millimeter. Obviously, this system has severe limitations. It can only be employed where the ground is hard and the sliding bars cause no disturbance of the soil crust. It may only be used on a small number of sites since transportation, establishment, and recording all take a relatively long time. Campbell, in fact, was only able to study 9 such meter squares for his studies in the Steveville Badlands of southern Alberta. Streeter (1975) has recently suggested that a simplified version of Campbell's equipment, consisting of a single bar fitted with sliding rods, would be useful in studies of foot-path erosion. The hard compacted surface of a footpath would be less liable to disturbance and penetration by the sliding rods than would be an undisturbed soil surface.

Literature Survey

The technique of erosion pins was pioneered by Schumm's studies of the Perth Amboy Claypit in 1952 (Schumm - 1956a). Miller and Leopold (1962) were the first to employ washers as an aid to recording, and the technique of Leopold and school has since become the most widely disseminated via the Vigil Network system (Emmett - 1965). The Vigil Network also includes some of the most protracted erosion studies. Leopold and Emmett (1972) have reported from sites where 9 years of data have been collected. Table 7.1 is a summary of the approaches and results of some of the more prominent and accessible erosion pin studies.

TABLE 7.1

THE USE OF EROSION PINS--LITERATURE SURVEY

Author	Location	Technique	Dates	Recording	Results and Conclusions
Colbert, E.H. (1956)	Cameron, Ariz. (Badlands)	Wooden stakes, exposed 153 mm set at selected sites.	June 1951-May 1955	2 recordings of 21 stakes.	Erosion: June 1951-July 1953 = 273 mm; July 1953-May 1955 = 111 mm. 6 stakes lost to vandals. No correlation of erosion and slope angle.
Schumm, S.A. (1956a)	New Jersey: Perth Amboy (Badlands developing on an infilled claypit)	Wooden stakes (464 mm x 6 mm) driven flush to the soil surface. Transects down slope orthogonals. (Stake separation 0. 3 m.)	June-Sept. 1952 (10 wks)	4 recordings of 16 transects on diversely oriented slopes.	Erosion: 11-39 mms. Parallel retreat of straight slope segments. No detectable changes in slope angle (1949-1952).
Schumm, S.A. (1956b)	South Dakota: Wall (Badlands)	Steel rods (464 mm x 13 mm) oriented normal to the surface and driven flush with the slope. Transects.	July 1953-Aug. 1955	2 recordings after 15½ months and 25½ months.	Erosion: 20-38 mms. Erosion depth increases with slope angle on wash controlled slopes (Brule Formation), but is at maximum on the convex divides of permeable slopes (Chadron Formation). Adjacent pediments retreated 24 mm in the same period.
Hadley, R.F. & S.A. Schumm (1961)	Nebraska: Sioux Co. (Badlands)	Steel rods (930 mm) driven flush to the soil surface. Transects down slopes and along divides.	June 1953-(June 1954) Oct. 1954	One recording of slopes after 16 months, of divides after 12 months.	Erosion: 8-38 mms (slopes); 15-46 mms (divides)

TABLE 7.1--Continued

Author	Location	Technique	Dates	Recording	Results and Conclusions
Schumm, S.A. & G.C. Lusby (1963)	West Colorado: Montrose & Badger Wash (Mancos shale terrain)	Steel rods (464 mm x 13 mm) oriented normal to the soil surface. Transects.	June 1958-April 1962	Semi-annual. 8 recordings of 25 transects.	Rods exposed least in spring and most in the fall, due to frost lifting of soil in winter and compaction by rainsplash in summer.
Leopold, L.B., Emmett, W.W. & R.M. Myrick (1966) Miller, J.P. & L.B. Leopold (1962)	New Mexico: Upper Arroyo de los Frijoles and Las Dos (Vigil site)*	Iron nails (254 mm), put through a washer and driven flush to the soil surface. Transects, channel cross-sections and grids. Washers employed to indicate cut and fill. Surveyed elevations of nail heads to demonstrate absence of frost disturbance.	1958 et seq. (1964) (1972?)	(?) Annual records of 202 erosion pins.	Erosion: Slope Wash Tributory Experiment (a) 23 mm p.a. (61 pins) (b) 44 mm p.a. (57 pins) (c) 2 mm p.a. (19 pins) Coyoto C. Arroyo: 73 mm p.a. (65 pins) No systematic trend in erosion quantities recorded. Pins on steeper slopes and nearer rills show a slightly higher rate of erosion.
Emmett, W.W. (1965) (1974)	Wyoming: Hudson, Last Day Gully (Vigil site)*	Iron nails (254 mm), with washer, set flush to the soil surface. Left one year to settle as a counter to disturbance caused during establishment. Washer returned to ground surface and nail reset after the third.	(1962) 1963-1968		Erosion: 3.3 mms p.a. (Standard Deviation +8.2 mms). Average erosion of 0.7 mms p.a. on slopes of circa 12 degrees.
Emmett, W.W. (1965)	New Mexico: Mittenrock (Vigil site)*	Iron nails (254 mm), with washer, set flush to the soil surface. Transects.		One annual reading from 25 nails in 2 transect lines	Erosion: 46 mms on a 20-degree slope.

Tinker, J.R. (1970) Clayton, L.K. & J.R. Tinker (1971)	North Dakota: Little Missouri (Badlands)	Iron pins (152 mm), with non-rustable washers used to indicate cut and fill. (78-mm iron pins tested but results discarded due to disturbance. Iron rods [620 mm, 1240 mm] without washers tested, but results discarded due to influences of cyclical changes in ground height.) Grids (and transects), also rods with washers established at seepage steps.	July 1967-July 1969 (Seepage Steps: April 1968-July 1969)	Results based on semi-annual records from 360 (155 mm) pins in 3 grids, and 5 recordings of 19 seepage step sites.	Retreat: West facing slopes, Sentinel butte formation: 9 mm p.a. S.W. facing slopes, Tongue River formation: 4 mm p.a. N.E. facing slopes, Tongue River formation: 3 mm p.a. Erosion perpendicular to seepage steps: 6 mm p.a. Slope wash contributed 99.9% of all sediment production.
Kirkby, A.V.T. & M.J. Kirkby (1974)	Arizona: E. Scatacon MIJ (Pediment)	Iron pins (254 mm x 10 mm) plus loose fitting washer to record cut and fill. Cross slope lines at 60, 250, 340 m below break of slope	Jan. 1964-Aug. 1965	2 recordings after 6 months and 12 months.	Mean erosion of Pediment at: 60 m downslope: -0.23 mm 340 m downslope: -0.12 mm Erosion channel washes: -8.1 mm Vegetation mounds: +5.2 mm Interfluves area: +0.6 mm
Bridges, E.M. (1969) Bridges, E.M. & D.M. Harding (1971)	South Wales: Lower Swansea Valley	Iron pins (152 mm) plus washer, allowed 10 mm exposure. Set at 0.5 m intervals across interfluve and down slope in a small gully system measuring N-S 120 m, E-W 15 m.		Weekly recording of 168 erosion pins.	Erosion: average 10.6 mm p.a. 5° slopes 10 mm p.a. 45° slopes 32 mm p.a. 62° slopes 43 mm p.a. Erosion maxima associated with puffing up of soil due to needle-ice formation. 19.5 mm lost between September and February from a 9° sloping interfluve.
Haigh, M.J. (1974)	South Wales: Blaenavon (Colliery spoil mounds & infilled opencast sites)	Steel pins (610 mm x 6 mm), with and without close fitting washer. Iron pins (152 mm) with and without loose fitting washer (removable washer used for measurement of pins with no permanent washer), allowed 14-18 mms exposure.	(April 1972) June 1972-June 1974	4 recordings each year of 587 erosion pins.	Erosion: Milfraen Spoil Tip Complex. Vegetated slopes: 3.1 mm p.a. Unvegetated slopes: 4.6 mm p.a. (June 1972-June 1974)

TABLE 7.1--Continued

Author	Location	Technique	Dates	Recording	Results and Conclusions
Haigh (cont.)		Established as 12 grids on downslope orthogonals.			
Harvey, A.M. (1974)	Westmorland: Howgill Fells	Iron nails (254 mm) driven through a washer and set flush to the ground at selected sites.	May 1971-May 1972	Recordings at 4-8 week intervals of 115 erosion pins.	Erosion: 0.8-6.4 mms
Imeson, A.C. (1971) (1974)	North Yorks. Moors: Hodge Beck	Brass welding rod (300-400 mms) established in quadrats.	1967-1969	Recordings of 800 erosion pins.	Retreat/advance under different vegetation types and on bare ground. Retreat: bare mineral soil 63.9 mm p.a. Bare peat 40.9 mm p.a. Advance: grass +2.5 mm p.a. Bracken: +6.6 mm p.a. *et alia*.
Pinczes, A. (1971)	Hungary: Tokaj Mountains loess (Vine-lands)	Iron stakes established on a 175 m x 50 m plot.	1961-1971	Not known.	Neutral surfaces and accumulation areas may occur on steep parts of the slope.
Temple, P.H. & D.H. Murray-Rust (1972)	Tanzania: Eastern Uluguru Mountains, Mfumbwe (Hill rice & regenerating bush)	Angle-iron rods (900 mm x 30 mm) graduated in inches. Rods sunk to half length in lines down the center of a 10 m cleared plot. Soil conservation measures applied.	March 1957-July 1960. Revived Feb. 1969-Aug. 1970	Circa annual on 8 plots with a total of 35 rods.	All cultivated plots showed heavy ground advance at the base of the slope 24-83 mm p.a. Maximum retreat on control plot (no soil conservation measured). Regenerating bush: advance at slope foot 1-25 mm p.a. Retreat: 1-17 mm p.a. Erosion only 60% of that during cultivation.

Diseker, E.C. (& E.C. Richardson) (1961) (1962) (& J.T. McGinnis) (1967) (& J.M. Sheridan) (1971)	Georgia: Cartersville (Artificially devegetated highway cuts)	Metal pins (9 mm thick) driven flush to the soil surface established in ditches and 3 rows in road banks. Ditch pins established in squares. Measurement by a special rod and bar device laid across tops of pins.	(Sept. 1956) Feb. 1958-1962	Annual recordings at 0.3 m intervals along squares and diagonals of pin quadrats.	Erosion: 23-53 mm p.a. on a 19.5° slope. Northwest facing plots lost more than twice the amount lost by plots facing southeast.
Campbell, I.A. (1970) (1970) (1973) (1974)	Alberta: Steveville (Badlands)	Movable contour plotting frame (a modified Schefferville Frost-heave Bedstead). Allows 25 measurements in a 25-meter square accurate to 0.5 mm, mounted on 4 modified erosion pins.	June 1969 et seq.	Circa annual on 9 plots.	Retreat Plot (1) -7.4 mm; (2) -4.5 mm; (3) -2.0 mm; (4) -17.4 mm; (5) -20.3 mm; (6) +5.8 mm; (7) -16.6 mm; (8) +2.5 mm; (9) -26.9 mm. Slope angle not as important as site, lithology & response to environmental fluctuations (e.g. freeze-thaw) in controlling rate of retreat.
Streeter, D. (1975)	S. England: Chalk Downland Footpaths	Movable contour plotting bar (modified version of Campbell's Frame) mounted on 2 modified erosion pins.	1974 et seq.	Not known.	Not known.

*N.B. There are 23 vigil sites in the United States employing similar techniques (Hadley - 1965).

Field Test of Erosion Pin Techniques

An attempt was made to conduct a comparative field examination of six basic approaches to the use of erosion pins (Figure 7.1). Two lengths of metal pin were employed: 600 mm (cf: Schumm - 1956 et seq.) and 150 mm (cf: Bridges - 1969; Bridges and Harding - 1971; Leopold et al. - 1966; Clayton and Tinker - 1972). Both types of rod had a radius of 5 mm. Additionally, two types of permanent washer and two types of removable washer were tested. Early experiments tested small (38 mm diameter), close fitting (6 mm gauge), permanent metal washers and a wide (66 mm diameter) loose fitting (15 mm gauge), removable plastic disc. Later experiments tested small (42 mm diameter), loose fitting (7 mm gauge) metal discs as both permanent and temporary washers. These field tests were conducted on 12 grids containing a total of 587 erosion pins which were established on two infilled opencast sites in South Wales and four coal spoil tip complexes in South Wales and North Staffordshire. It was discovered that both accuracy and performance differed amongst the various erosion pin types.

Replication of Results

The most obvious difference between the different erosion pin methods tested was the degree to which an individual set of readings could be replicated. Table 7.2 displays the results of repeated field checks undertaken for each type of erosion pin. All checks were made in fine weather and it might be expected that adverse weather conditions would create the possibility of higher margins of error. It can be seen that the pins with the permanent washers give results which are most closely duplicatable, and that the larger and removable washers give results which are progressively more and more variable. There was no significant difference between the performances of the two lengths of erosion pin during this experiment.

Technique Employed in South Wales Studies

The experience gained from these early field tests was employed in the design of the erosion pin methodology adopted at Blaenavon. It was decided to use erosion pins of both 610 mm and 153 mm lengths. There still persisted some doubts about whether the shorter pins would be affected by soil creep so it was decided to locate 153 mm pins adjacent to the longer 610 mm pins which might be expected to be less affected by disturbance. After some early experi-

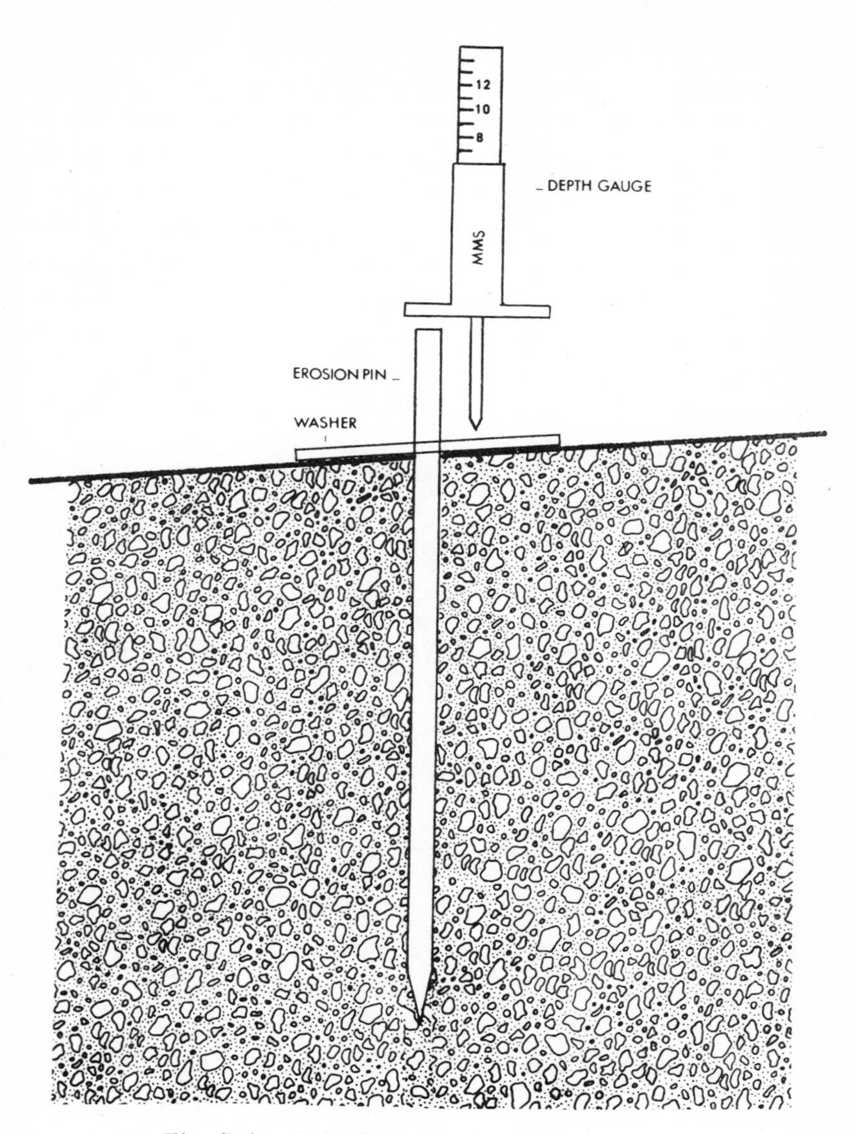

Fig. 7.1.--Data Collection from Erosion Pin

TABLE 7.2

MEAN DEVIATION OF REPLICATED DATA MEASUREMENTS FROM INDIVIDUAL EROSION PINS WITH DIFFERENT TYPES OF WASHER

Washer Specification	Mean Deviation
Close fitting, permanent, metal (38 mm dia.)	0.47 mm
Loose fitting, permanent, metal (42 mm dia.)	0.53 mm
Narrow, removable, metal (42 mm dia.)	0.68 mm
Wide, removable, plastic (66 mm dia.)	1.26 mm

ments this policy was formalized in the design of a standard format erosion pin grid row. The standard format row was one meter wide and bounded at either extreme by a 610 mm rod. Three 153 mm pins were established at 250 mm intervals between such that the shafts of all five pins lay in a vertical plane. It was decided that it was unwise to set the erosion pins flush to the ground surface because of the risk of burial. Similarly, it was considered that a large exposure might tend to make the pins a more obvious target for vandals. A compromise exposure of circa 14 mms was adopted. However, it was also considered advisable to determine whether the erosion pins suffered vertical displacement during the experiment. In some locations it was possible to arrange the pins in a standard format row such that the tops of all the pins lay in a horizontal plane. In such locations rather larger exposures were allowed to certain pins.

It was decided to employ a permanent close fitting washer through most of these experiments because this was the system which gave the most replicable data results (Table 7.2). The only experiment to adopt a different system was Waunafon B. This grid was established before the replication experiments and data was collected via a large removable washer (Table 7.2).

Disposition of Erosion Pins upon the Slope

The literature survey (Table 7.1) reveals that there are two approaches to the arrangement of erosion pins on the hill-slope. Some authorities prefer to cluster the pins at sites which they consider are likely to yield especially interesting results, others prefer to distribute the erosion pins as regular grids or transects. Once again, this study adopts a compromise approach. In most of the experiments, standard format grid rows are laid out as transects up the line of steepest slope. The spacings of the rows are based on multiples of a 1.5 m unit and the most usual spacing is 3 m. However, a more dense spacing of the rows is adopted on sections of the slope where there is a more rapid change in slope angle than on the more rectilinear slope units.

In some situations there has been an attempt to monitor slope retreat in association with activity in a parallel gully channel. This has been accommodated in two ways: by increasing the number of erosion pins in the grid row, and by altering the internal spacings of the grid row. At Waun Hoscyn, certain standard grid rows have been expanded by the addition of extra 153 mm and 610 mm pins at the appropriate distances. At Milfraen, one grid has been designed as a double grid with an internal spacing multiplied by a factor of three. At Waunafon, B/C grids comprise a double grid with an increased number of 153 mm pins and a lightly larger internal spacing. This grid again has a lateral spread of 3 m.

Specific details of the numbers, distribution, performance, and problems encountered in the use of erosion pins on individual profiles is included with the discussions of results from each site. These discussions are contained in Chapters 11-13 and 15-17.

Seven Sources of Data Contamination

The behavior of the erosion pins was observed over a period of two to three years. It proved possible to establish that there were seven main sources of data contamination. These were:

(i) disturbance during establishment;

(ii) disturbance of the pattern of soil erosion caused by the presence of the erosion pin;

(iii) disturbance of the erosion pins caused by the physical differences between them and the surrounding soil;

(iv) disturbance of the erosion pin by trampling and vandalism;

(v) the effects of variations in the erosion pin's environment;

(vi) disturbance of the erosion pin by the operator during recording;

(vii) errors in recording.

It was found that different categories of erosion pin were subject to different types of data contamination to different degrees.

(i) Disturbance during Establishment

It is impossible to establish erosion pins on a slope without causing a considerable disturbance of the soil surface. This disturbance is especially marked when that slope is steep, unstable and unvegetated, and when the erosion pins are long and require a great deal of hammering into the ground. Shorter pins, like those used by Leopold (op. cit.) and Bridges (1969), are relatively easy to push into the ground and their establishment creates far less disturbance.

It has been suggested that one method of avoiding the effects of initial disturbance might be to allow the pins a period of grace between establishment and the collection of the first data record. Emmett (1965) recommends that this period should be as long as one year. In this study, the author allowed one to three months between establishment and initial data collection. In subsequent studies in Arizona (Haigh - 1977: In preparation) recording was delayed until after the first period of heavy rainfall.

(ii) Disturbance of the Pattern of Soil Erosion Caused by the Presence of the Erosion Pin

The presence of the erosion pin disturbs natural soil erosion in two ways. Firstly, the pin itself impedes soil movement. Stones become held on the upslope side of the pin causing an accumulation matched by the development of a hollow on the downslope side due to insufficient replacement of eroded material. This effect is especially noticeable in the case of pins established on steeper and stonier slopes and in rill and gully channels. In the latter case, sometimes, there is evidence of accelerated erosion at the base of the pin due to the interference with flow. Secondly, the presence of washers causes problems because of their tendency to protect the underlying soil from rainsplash. In summer particularly, washers may be found standing up to 5 mms above the level of the surrounding soil. Washers may also interfere with soil movement and sheet flood, sometimes to a greater extent than does the erosion pin itself.

(iii) Disturbance of the Erosion Pins Caused by the Physical Differences between Them and the Surrounding Soil

Erosion pins may also serve to contaminate their own data. This is particularly true of the systems which employ shorter pins and washers. Erosion pins tend to be of a different density to the surrounding soil. Unless they are sufficiently firmly anchored (e.g., by rust), or deeply buried, the natural vibration of the soil will cause differential movements of the markers in the soil, particularly settling.

Metal erosion pins tend to have a better conductivity than the surrounding soil. They tend, therefore, to be favored sites for the growth of ice crystals. Similarly, the undersides of washers are frequently the site of the growth of needle-ice. Frost heaving is most noticeable as an influence on shorter pins with tight fitting or rustable washers.

(iv) Disturbance of the Erosion Pin by Trampling and Vandalism

Trampling and vandalism are two major sources of data loss. Erosion pins fitted with washers are more obtrusive, hence more liable to discovery by children or vandals. Longer pins are relatively immune both from treading and vandalism. Once established they are very difficult to move.

(v) The Effects of Variations in the Erosion Pin's Environment

The erosion pin record is complicated by rhythmic changes in the natural environment: wetting/drying, freezing/thawing, together with the annual round of vegetation growth and decay; all affect the height of the local ground surface. Schumm and Lusby (1963) have noted that stakes show minimum exposure in spring when the soil is puffed up by needle-ice formation, and that the soil is beaten down and compacted by rainsplash during the summer. These observations were verified by this case study. Sometimes the needle-ice effect is so great as to temporarily dwarf the effect of erosion. The recording of a mean ground advance across a whole slope is not an unprecedented discovery. A consequence of the activity of these forces is that even a totally neutral slope, with no erosion or deposition, may give rise to quite a complicated record of small scale ground advance and retreat.

(vi) Disturbance of the Erosion Pin by the Operator during Recording

Just as the establishment of erosion pins causes erosion, so does the recording (consider Photograph P.7.1). The operator disturbs the soil by gaining access to the erosion pin and he inevitably disturbs the erosion pin when he records its exposure. Disturbance of the soil is an increasing problem as the slope increases in steepness. On unvegetated slopes it is virtually impossible

P. 7.1. --Disturbance of 153 mm Erosion Pins on a Steep-sloped Site Affected by Soil Creep and Some Trampling.
(Photograph: June 1975.)

to avoid rupturing the soil crust or disrupting the stone layer. Disturbance of the pins is especially likely when the pin is short and when it is fitted with a washer. Washers generally have to be pushed down to the soil surface, rust bonds have to be broken, accumulations due to rainsplash protection ground down. The more frequently recording is undertaken, the more disturbance caused.

(vii) Errors in Recording

Even the most rapidly eroding hillslopes suffer ground retreat which can only be counted in terms of a few millimeters each year. It is very difficult to gather recordings which are accurate to the nearest millimeter. It is imperative that one know the range of accuracy of the recording method employed. It is also imperative that the results recorded should be far greater than the range of experimental error.

In hindsight, the author concedes that the choice of materials, design and layout of the experiments described in this volume might well have been improved. For the benefit of those who follow, therefore, the author would like to propose the following guide-lines.

Recommendations

(i) Erosion pins should be thin metal rods with a length of not less than 600 mm. Shorter pins are unsatisfactory because they are too easily disturbed by frost-heave, trampling, and soil creep. However, the establishment of shorter pins causes less soil disturbance and their use may be appropriate to short, summer-time, studies of areas suffering particularly rapid erosion.

(ii) Permanent washers should be loose fitting and non-rustable. Removable washers should be as close fitting as possible.

(iii) Washers are not considered to be entirely suitable for the measurement of cut and fill. They are poor sediment traps, they cause local disturbance of soil movement and sheet-flood; further, resetting after each data record creates an unacceptable amount of surface disruption.

(iv) Erosion pins should not be set flush to the ground surface lest they become buried and are lost. Deposition may affect almost any section of any slope at any time. An exposure of 15-20 mms seems reasonable.

(v) The detail of ground retreat is very complicated. Readings are likely to give a better general picture if they are the result of the multiple remeasurement of a number of different pins established at equivalent sites. Erosion pins are best established in short rows or small clusters. If the heads of all the pins in a particular row or cluster are aligned in a horizontal plane, then it is possible to detect pins which have suffered minor disturbance, and eliminate their results from the data record.

(vi) Erosion pins should be recorded semi-annually. Increased frequencies of recording may be appropriate for more stable, vegetated slopes. However, the operator should be careful to avoid causing too much of his own erosion.

(vii) Studies employing erosion pins might profitably be supplemented by the collection of the following data: (a) rainfall: amount and intensity, (b) ground temperature--number of freeze-thaw cycles, (c) soil moisture variations, (d) vegetation: cover, and type, root depth and root density, (e) land use, numbers of livestock, distribution of trampling, (f) mechanical properties of the soil, (g) rate of soil creep.

CHAPTER 8

TECHNIQUES II:

PHOTOGRAPHIC AND MORPHOMETRIC

The main studies contained by this thesis are concerned with the use of erosion pins in the direct measurement of morphometric change. However, several other techniques have been employed during this investigation to complement and to supplement these results.

Erosion pins measure retreat at a single point. They provide little information concerning the movements of the spoil in which they are situated and are unlikely to yield a good record of some of the more localized, irregularly distributed, and seasonal phenomena, which affect patterns of ground retreat. It was decided, therefore, to initiate a series of experiments based on the sequential photography of permanent quadrats with the object of obtaining a detailed record of surface spoil movements and gaining an insight into their effects on the ground retreat data records.

Nine permanent quadrats were established. Seven are located at Cefn Garn-yr-erw on the Waunafon Opencast and two on the northern slopes of the plateau spoil tip complex at Milfraen. Each quadrat consists of a meter square bounded by four (610 mm) erosion pins and a permanent camera station which is marked by a fifth pin (Photograph P. 8.1).

The photographs are taken by a "Practica" single-lens reflex camera tripod mounted at constant height and inclination above the fixed camera station. During photography, the quadrat is emphasized by a linen tape, graduated in 100 mm units, which is looped over each marker to define the quadrat's perimeter. The quadrats are re-photographed at three monthly intervals which coincide with the quarterly remeasurements of the main erosion pin experiments.

Photographic quadrats were established at selected sites of special interest. Their subject matter includes the performance of erosion pins and the migration and dispersal of lines of marked stones, the effect of an inflexion on rates of spoil movement, the erosion of a tuft of residual vegetation, the effects of sheep trampling, rill and gully development.

P. 8. 1. --Sequential Photography of Permanent Quadrats

These photographs were further supplemented by a series of more general photographs. These were based on the repeated photography from a fixed camera station of the area around a single fixed target. These photographs, which, again, were taken at three monthly intervals, include an attempt to study the evolution of rills and gullies (Photograph P. 8. 2).

The results from the photographic experiments are not discussed separately in this study. However, these photographs have been widely incorporated in the general text to illustrate and to qualify results from the erosion pin grids.

Some further attempts were made to record the movements of the spoil's surface layers. Buried columns of sand or calcite crystals were employed in two separate attempts to record a soil creep profile. However, it proved impossible to re-excavate these columns sufficiently cleanly to gain any useful results. Wash traps were also established adjacent to two erosion grids: Waunafon C. and Waunafon W. However, it proved impossible to disguise these traps sufficiently to avoid repeated vandalism. Further, the results gathered were highly anomalous and evidently owed more to disturbance during establishment than to the normal processes of erosion. Fifteen wash traps were established and results were gathered for six months. At the end of that period only two traps remained undisturbed. It was decided to abandon the experiment.

Slope morphometry forms an important part of this study. Detailed and accurate morphometric measurements were required for each of the instrumented slopes to provide an accurate baseline for the description of the patterns of slope evolution suggested by the measurements of ground retreat. Morphometric analyses were also employed to extend the scope of this study to slopes which could not be instrumented.

Slope profiles were recorded by means of a one meter slope pantometer (cf. Pitty - 1968). It has been suggested that unit-length slope profiles provide a more objective description of slope morphology (Pitty - 1967). It has also been demonstrated that, because of the influences of micro-relief features on the variability of slope angle records, results from measurements based on different unit lengths should not be statistically compared (Gerrard and Robinson - 1971). All slope data employed by this study are based on measurements of a one-meter unit length.

The accuracy of the slope pantometer was the subject of a recent paper by Morris (1974). Morris discovered that the mean deviation of the results gathered by his 1. 51 m slope pantometer was 0. 2° on slopes of less than 17°, but

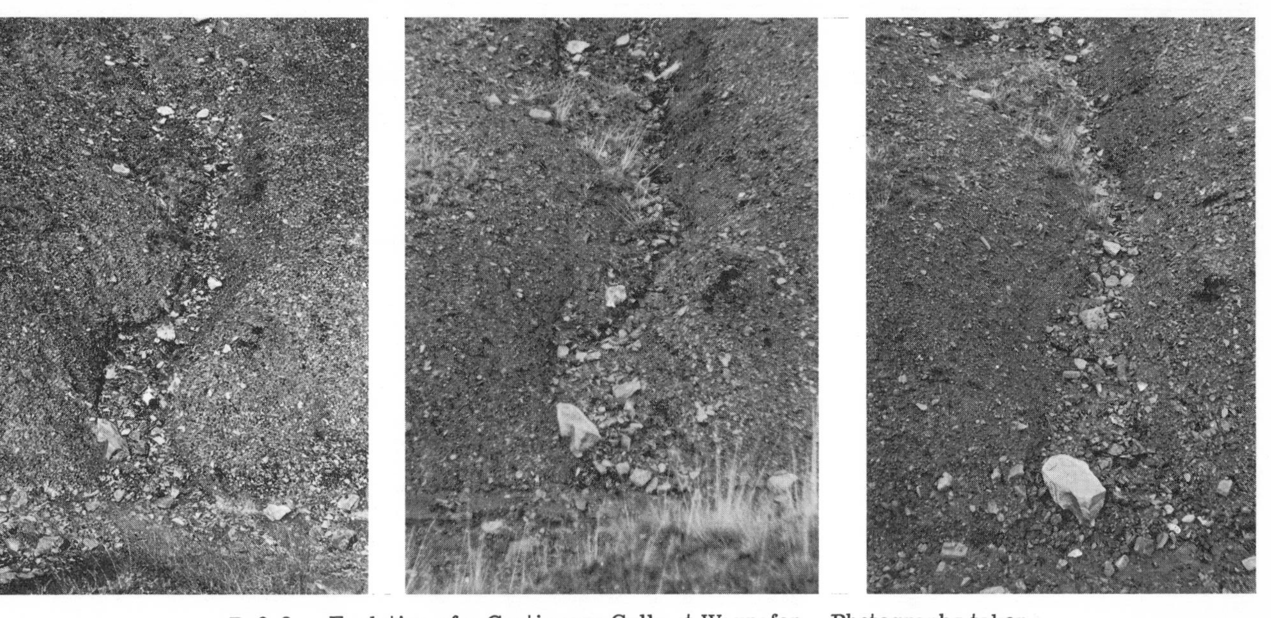

P. 8. 2. --Evolution of a Continuous Gully at Waunafon. Photographs taken in October 1972, 1973, 1974.

0.5° on slopes steeper than 17.5°. Field tests have been undertaken to determine the accuracy of the instrument employed by the present study on a carefully monumented slope profile whose slope angles ranged from 15° to 21°. The mean deviation of pantometer records about these mean readings was 0.4°, the standard deviation was 0.6°. The total range of error was 1.6°.

All the slope profiles recorded in this study are based on the measurement of the flowline slope. This is the steepest slope which can be recorded at any given site. The concept corresponds with the geologists' idea of the "true dip" of rock strata. Certain of the flowline profiles employed to illustrate the results from the erosion grid experiments are not true flow-line slope profiles but based on the flow-line slopes to a distance of one meter above or below any particular row of instrumentation. These local flow-line slopes are scaled to represent the actual distances separating the erosion grid rows.

An attempt has been made throughout this study to record parallel slope and gully profiles. These records are supplemented by measurements of gully incision. Gully incision, here, is the vertical distance separating the floor of the gully channel from the up-slope point of each meter-unit measured on an adjacent interfluve flow-line profile.

Once again, most of the results from these studies are incorporated into the chapters which discuss the results from the erosion grids. However, Chapter 10 is devoted solely to morphometric analysis of the slopes of Blaenavon's nineteenth century "fan-ridge" coal-spoil tips.

CHAPTER 9

THE COLLIERY SPOIL MOUND ENVIRONMENT

(REVIEW)

The production of one ton of clean coal by underground mining creates 0.5 tons of spoil. During the past 100 years approximately 3,000 million tons of coal have been lifted from the South Wales Coalfield (Jones et al. - 1971). The current annual production is 18.6 million tons (Morgan - 1965), and the National Plan of the N.C.B. (N.C.B. - 1950) puts estimated reserves at circa 5,000 million tons. One concludes that about 1,500 million tons of coal-spoil has been produced in the South Wales Coalfield, that production continues at more than 9 million tons p.a., and that that total of spoil produced will be at least doubled before the South Wales coal mining industry disappears. In 1966, 7,000 acres in South Wales were buried beneath coal-spoil (Thomas - 1966). That area must surely increase and so it becomes important to study coal-spoil as a landforming material (Yatsu - 1970), and to understand its modes of morphological evolution.

Mineralogical Composition

Colliery spoil mounds are highly variable in composition. Some, especially the older tips, may contain as much as 25% of carbonaceous material. However the bulk of the tip is usually composed of a stable argillaceous shale. Mineralogically, 90% of the non-carbonaceous material is likely to be a mixture of quartz, kaolin, illite, and various weathering compounds (Figure 9.1).

Particle Size

There are two basic types of coal-spoil: 1) coarse discard which is mainly "run-of-mine" waste, 2) fine slurry or "tailings." Sixty million tons of spoil were produced in 1970, of which 55 million tons were coarse discard and 5 million were fine discard. The coarse discard is mainly larger than 5 mm in size. Jones et al. (1972) state that it might be expected that of such material, 90% would pass through a 508 mm mesh, and 90% would be retained on a 0.75

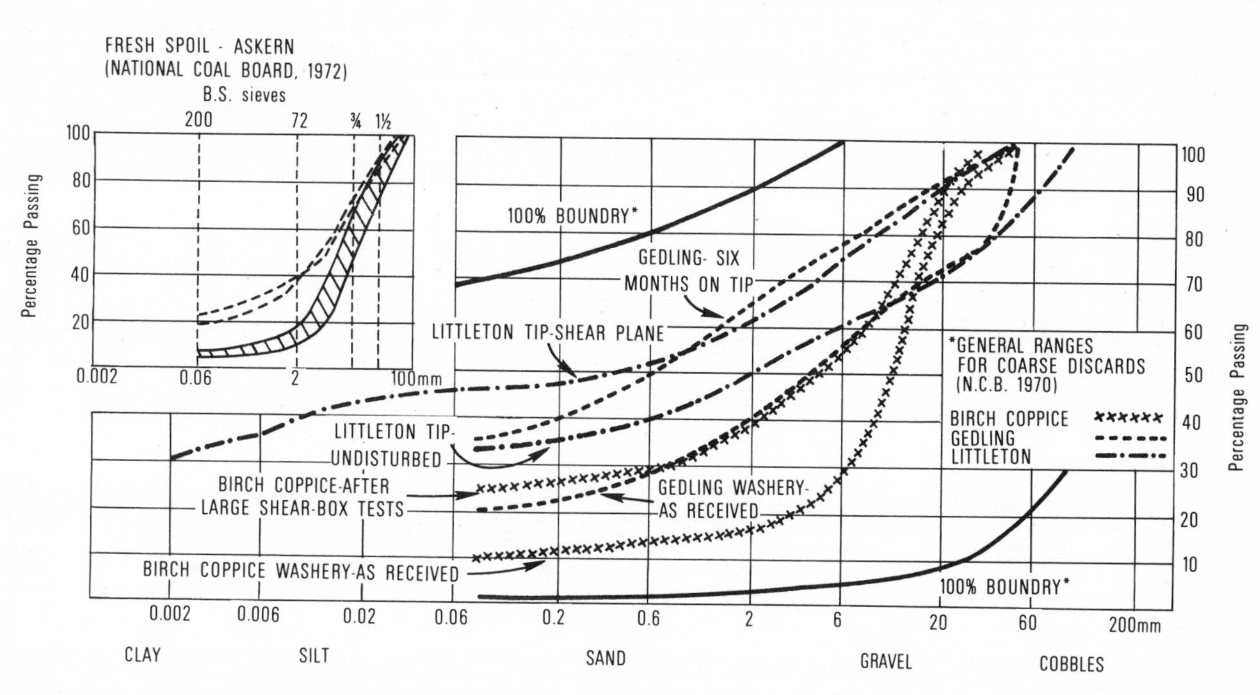

Fig. 9.1.--Particle Size Distribution Comparisons: (i) Mechanical breakdown after transportation and spreading--Askern (insert: NCB, 1972), (ii) Breakdown after 6 months exposure--Gedling, (iii) In situ shear plane and undisturbed discard close to the shear plane--Littleton, (iv) After large shear-box tests (pre-loading technique; 2.5 m displacement) on fresh washery discard--Birch Coppice (from Taylor - 1976).

mm mesh (B.S. 200). Variations in particle size within the tip are mainly due to segregation in the tipping process.

Fine discard is also of two types: slurry and tailings. Both are composed of particles smaller than 0.5 mm. Slurry is a mixture of coal and spoil which has not been cleaned. Tailings are the waste left after fine coal has been separated by froth flotation in a stilling tank (Smith - 1968).

Fine discard is usually dumped separately. Such spoil tips are dense, coherent, and may possess a surface crust of some tens of millimeters thickness. These tips are structureless and entirely devoid of vegetation (Hall - 1957).

Particle Specific Gravity

The density of a spoil tip is determined by its particle specific gravity. This can vary from as little as 2.0 in an unburnt tip with a high coal content to 3.0 in a burnt-out tip. It may also vary with particle size. Particle sizes below 190.5 mm have on average a specific gravity of *circa* 2.35.

Bulk Density

The *in situ* bulk density on an unburnt South Wales coarse discard spoil tip is estimated as being between 1.28 and 2.40 gm/cc (Jones *et al*. - 1972). Standard compaction tests carried out on spoil tip materials showed that the low value for maximum dry density of 1.73 gm/cc was relatable to the low particle specific gravity in a loose-tipped spoil heap. Sand replacement tests undertaken at a depth of 300 mm indicated a dry density of the order of 1.31 gm/cc. It was calculated that if the material were compressed, a relative compaction of 100% might be achieved. The spoil heap would then occupy 25% less volume (Smith - 1968a).

Creation and Stability

Colliery spoil tips are generally created by one of two methods. Either they are loose-tipped from a MacClane Shute or Aerial Bucketway, or they are spread in layers and compacted by tractors. Loose-tipping creates a lightly packed tip whose slope-angles are controlled by the material's angle of internal friction. They are marginally stable but they may become more stable as the dumping of more material causes consolidation in the lower layers. Most old tips were compacted inadvertently by tipping from tramways laid across their upper surfaces. Since the Aberfan disaster there has been a return to compacted tipping. Most civil engineers prefer compacted tips because they are stronger, less liable to subsidence, space-saving and less liable to combustion. Smith (1965b), however, considers that loose tips may suffer less disruption by settlement than compacted tips. His studies show that compacted spoil loses strength rapidly once peak strength is exceeded. The loss, which occurs over a small strain, may reduce its strength to a level comparable with the maximum for loose-packed spoil. Loose compacted spoil maintains a constant sheer strength value once its maximum has been reached. Peak shear strengths of spoil materials vary with variations in local lithology and percent coal content, and often differ amongst individual tips in a single area. Thomson and Rodin (1972) suggest that the lower limit of peak shear strength is 25° and the upper limit just over 40°.

The question of spoil-tip stability has been the subject of a great deal of research. Traditional geotechnical analyses of coal tips and their stability are discussed by Smith (1968b), Thomson and Rodin (1972ab), Jones _et al_. (1971), Spears _et al_. (1971, 1972), and others.

Smalley (1972) discusses the role of inter-particle bonding in influencing flow-slides in fine-particle mine waste. It is considered that the presence of particles with low settling velocities in water, and a high moisture content indicates that the material is dominated by short-range inter-particle bonds. Short-range bonding is dangerous because a small disruption of structure can cause a great loss of strength (_cf_. quick clays). The problem disappears if active clay particles are present since these contribute long-range bonds to the system. Long-range bonds can accommodate deformation (Smith - 1968b). Bishop extends discussion to a post-rupture history of slides which is related to a brittleness index (Bishop - 1967). It is emphasized that waste materials, which in their loose state are most liable to flow-slide, are in general excellent construction materials when compacted at the correct water content. Chwastek (1970) has studied the effect of weathering on slope stability in Polish lignite mine dumps. Weathering was found to reduce the degree of inter-particle bonding and halve the angle of internal friction.

Relationship between Slope Height and Slope Angle

Studies of the effect of mining procedure and geological structure on Polish lignite mine dumps at Patnow have discovered a relationship between the height of the dump's scarp and its general angle of inclination:

$$a = 0.0129\ h + 0.739$$

where h is the height of the scarp in meters and a is the general angle. The general angle is defined as the angle subtended by a line connecting the highest and lowest points of the slope profile and the horizontal (Chwastek - 1970). The strength of this correlation was 0.692.

The Importance of Compaction to Slope Development

Early studies of the processes of erosion and slope development considered that an increase in slope angle automatically resulted in an increase in soil loss (Duley and Hays - 1932; Borst and Woodburn - 1948; Meyer and Monke - 1965). Recent studies, however, have shown that this need not necessarily be the case and that the patterns of erosion and slope development are strongly influenced by the degree of compaction (i.e. "unit weight") of the soil materials (Foster and Martin - 1969; Rowlinson and Martin - 1971). It has been discovered that for materials which have a low unit weight, erosion rates are greater on the less steep slopes, while for materials with a high unit weight steeper slopes suffer the greater erosion.

Physical Weathering

Exhumation causes a dramatic alteration in the physico-chemical environment of coal-spoil materials. The most important changes are the result of contact with water. The alternation of wetting and drying in exposed coal-spoil materials causes a rapid break-down of the shale. During dry weather, evaporation from the rock's surfaces promotes high suction pressures within the internal voids and these tend to become mainly air filled. Subsequent immersion causes this air to be pressurized as water is drawn into the outer pores by capillary suction. The build up of internal stress causes the rock to fail along internal planes of weakness which may be both macro-structural (bedding planes, laminations, joints) and micro-structural (the weakest bond in the mineral skeleton). Progressive failure within the rock opens up a progressively greater surface area for attack by the same processes. Taylor (1974) has reported that some weak shales may be reduced to sand-sized particles in a matter of weeks by such processes.

Profile Development

The development of a soil profile on colliery spoil materials is a slow process. If the shale contains no pyrites, the only profile development is by textural differentiation. The result is a simple situation in which the surface horizons are composed mainly of fine particles while lower levels contain progressively larger fragments (Table 9.1). The depth of weathering is a function of the age of the tip, the nature of the spoil materials, and the local climatic regime.

TABLE 9.1

RELATIONSHIP BETWEEN DEPTH AND PERCENTAGE COMPOSITION OF FINE PARTICLES AFTER 10 YEARS WEATHERING
(Doubleday - 1972)

Depth (mms)	% Less than 2 mms
0-100	71.8
100-220	75.3
220-370	15.5
370-520	15.9
520-880 plus	17.4

In Table 9.1 the reduced percentage of fines in the uppermost layers is probably a result of surface erosion. The processes of deflation rainsplash and sheet-flood might be expected to remove fines and leave behind coarser materials, sometimes as a surface stone layer.

Erosion and Slope Development

There has been very little research into the nature of the erosion of colliery spoil mounds. Repelewska (1968) has described the erosion of some waste dumps produced by the phosphorous mines at Annopol on the Vistula, Poland. Here, deflation was a very important process. Quartz particles of 0.5-0.25 mm were blown from the peak parts of the spoil dumps and redeposited on the slopes as the wind sculptured the tip surfaces into asymmetrical horizontal grooves and garlands. This aeolian activity was most noticeable in the summer. Gullying was severe (vide also Mizutani - 1970), and frequently led to the development of slope failures. The author's observations of the operation of this process in the Blaenavon area bring to mind the studies of Dauksza and Koratuba (1972). This work was an attempt to extend Thornes (1971), who applied queueing theory to the analysis of scree slopes, and adopted a similar approach to analyses of the development of fluvially undercut slopes by landsliding. Simply, the undercutting of the study slope creates a condition of instability and instigates landsliding. Landsliding supplies debris to the river which must be removed before the landslide is reactivated.

Sheet-flood and seasonal solifluction were also considered by Repelewska (1968) to be important to slope development. Water erosion was most important in the late spring and early summer. Seasonal solifluction was mainly associated with spring snow-melts and winter thaws.

Chemical Weathering

Fresh spoil is usually alkaline, with a pH of 7.1-8.4. However, as spoil weathers it becomes more acidic. This effect first becomes noticeable in the surface horizons of young spoil tips (Table 9.2), but tips which are fifty years old may be thoroughly acidic (pH 4.5-5.5), while tips which are a century old or more may yield pH values of 3.0 or less. Frequently, this increasing acidity is most apparent on upper slope sites. Hall (1957) gives these data from the Old Sewerage Pit Spoil Tip, Somerset: top of mound--pH 4.2; upper slope--pH 4.8; middle slopes--pH 4.9; lower slopes--pH 5.4.

TABLE 9.2

pH VALUE PROFILE IN A 10-YEAR OLD COLLIERY SHALE TIP
(Doubleday - 1969)

Depth (mms)	pH in Water (1: 1.25 suspension)
0-100	3.7
100-220	2.5
220-370	4.4
370-520	5.7
520-880	7.4
10 meters	7.5

The development of this acidity is due to the oxidation of iron pyrites:

$$Fe\ S_2\ (S) + 7/2\ O_2 + H_2O\ /\ 2\ SO_4^{--} + 2H^+ + Fe^{++}$$

$$Fe^{++} + \tfrac{1}{4}\ O_2 + 5/2\ H_2O\ /\ Fe\ (OH)_3\ (S) + 2H^+$$

Sometimes ferrous iron is present in soluble form in the spoil. This acts as a buffer. Acids produced by the pyrites may cause additional ferrous iron in soluble form to be released from the cementation of mudstones and siltstones (Palmer - 1969). When spoils have a neutral or stable pH due to the presence of carbonates, the weathering of pyrites will lead to the formation of amorphous ferric hydroxide and eventually the mineral goethite (Doubleday - 1972). Thus a mineralogical profile may be built up which is largely independent of the textural profile. It is interesting to note, however, that the studies of Spears et al. (1971) of a spoil heap at Yorkshire Main Colliery suggest that chemical weathering per se does not, in general, increase with depth into the spoil heap.

So far, these discussions of both the physical and chemical weathering of coal-spoil have been limited to the results of British studies. However, the results reported from other parts of the world are much the same. Geyer and Rogers (1972) demonstrate that after twenty years exposure, both the pH and the percentage composition of soil-size particles in unburnt Kansas spoil banks had increased. Pennsylvania's spoil banks were found to be highly acidic with a mean pH of 3.3 (Thompson and Hutnik - 1971).

Spoil Bank Fertility

Colliery spoil tends not to be particularly fertile either before or after weathering. Thompson and Hutnik (1971) discovered that none of the major plant nutrients were present in plant-available form in the spoil banks of Pennsylvania. Nitrogen was available in fossil form. Potassium was recorded in a few samples. However, deficiencies of phosphorous, calcium and magnesium were ubiquitous, and the spoil frequently contained toxic quantities of iron and aluminum.

In the spoils of the United Kingdom, the situation is very similar. Thus while most spoil tips contain a fair amount of potassium, the amount of this chemical which is present in plant-available form is very variable and often very small. The clay mineral illite and the basic sulphate jarosite ($KFe_3[OH]_6[SO_4]_2$) are the most usual hosts of the plant-available potassium. Some spoils have the capacity to render potassium fertilizers unavailable by incorporating the chemical within these structures. In one extreme case the potassium equivalent of an application of 15.15.15 fertilizer applied at 2 tons per acre was made unavailable within a single month (Palmer - 1969).

The phosphorous content of coal-spoil tends always to be low. Conventional studies of plant-available phosphorous usually yield a zero result (Palmer - 1969). There is a tendency for phosphorous content of spoil banks to decline with increasing time of exposure (Geyer and Rogers - 1972).

Burning

Spoil combustion has an important effect on spoil fertility. Ignition reduces shale bulk but improves fertility by residual enrichment. It also causes the breakdown of the clay minerals which are the main hosts of spoil potassium. The extent of this breakdown increases progressively with the temperature of combustion. The result is an important increase in plant-available potassium and a far more fertile spoil.

In Britain in 1967, 65% of the colliery spoil mounds owned by the N. C. B. were classified as "burnt out" or "burning." The mechanism of spontaneous combustion is imperfectly understood. Coal exposed to air has the capacity to absorb oxygen. One cc of absorbed oxygen generates 4.4 calories of heat in the Oyubari Lower Coal of Japan. Why this should occur is not known but pyrite decomposition is not the prime cause even though this mineral is strongly exothermic (Gunney - 1968). Nevertheless, self heating does occur and if the coal is fresh and if there is not an adequate flow of air to cool it, heat accumulates and the coal ignites (Hashimoto - 1974).

During combustion the shales suffer a significant loss of weight due to the oxidation of carbon and loss of structural water. The latter effect has often been noted and has given rise to a legend that damp is a cause of spontaneous combustion. Weight reductions of nearly 30% have been recorded in laboratory studies of combustion at 800°C (Doubleday - 1972). Burnt spoil lacks the coherence of unburnt shale. It is reduced to a heterogeneous mixture composed of a pink, flour-like dust with intermingled brittle shale fragments. There is a tendency for burnt shale below a depth of *circa* 300 mm to be fused as an amorphous or laminated mass. Sometimes these hard fused surfaces may be exposed by erosion. Burnt spoil has a more open texture, is more permeable, more easily leached, and weathers faster to a greater depth than unburnt spoil (Photograph P. 9. 1). It also tends to resist gullying more successfully than unburnt spoil (Hall - 1957).

Vegetation and Ruderal Colonization

The colliery spoil mound is a rather harsh type of ruderal environment. Its vegetation is controlled by the chemical properties of the spoil, its degree of comminution, the surface stability of the spoil, and the availability of suit-

P. 9.1. --Open Textured Burnt Spoil

able seed material in the surrounding landscape.

Several studies have considered the nature of the plants which inhabit colliery spoil mounds and the changes in plant types which occur during succession. Brierly (1956) found that 58% of all angiosperm individuals discovered on spoil tips were wind dispersed and that this included 36% of all recorded plant species. The Compositae were the best represented and this family alone accounted for one-third of all recorded plants. Most of the common composites were dispersed by seed parachute. Twenty percent of the flowering herbs were annual, 12% were biennial, and 68% were perennial. Perennials were more common on the older spoil tips. Down (1973) found that on Somerset coal spoil heaps 68-79% of the species were hemicryptophytes. On a twelve-year-old tip, rosette hemicryptophytes comprised 32% of all species, but on a 98-year-old tip only 12% were rosette hemicryptophytes.

Richardson (1958) puts forward the contention that there is no single pattern of plant colonization and species succession in colliery spoil tip communities. However, successions which are appropriate to single tips or small mining areas have often been described in the literature (Whyte and Sisam - 1948; Warwick - 1958). Rees (1955) has produced one of the more comprehensive descriptions of plant colonization in the Black Country on the three main types of colliery spoil mound. Table 9.3 summarizes the succession observed on 40-year-old low, probably fan-ridge type, hummocks.

Rees' discussions also appreciate the differences in the vegetation of the steep 20-40° slopes of plateau spoil mounds and that of the flat plateau summit (Table 9.4). In practice, however, these plateau summits support a richer and denser vegetation than is suggested by Rees' study.

MacClane Shute cones are the least hospitable of the spoil tip plant habitats, and those composed of fine discard tend to be entirely devegetated (Hall - 1957). Coarse discard cones tend to support a thin vegetation with poor cover and a very small species list (Table 9.5).

Burnt shale tips are more fertile and tend to be well vegetated (Hall - 1957). There are no known studies of species or succession.

Ecology

The detail of spoil tip configuration plays an important role in the ecology of its vegetation. Steeper slopes tend to support a vegetation which is less dense and contains fewer species than on shallow slopes because of spoil movement and surficial erosion. Similarly, concave slope embayments and slopes

TABLE 9.3

SUCCESSION ON LOW COAL-MEASURE SPOIL MOUNDS IN THE BLACK COUNTRY
(after Rees - 1955)

Stage	Species	
Stage 1 Soil pH 3.4	Agrostis tenuis Festuca ovina (High clay content sites:	Rumex acetosella Tussilago farfara)
Stage 2 Soil pH 3.6	Stage 1 species with:	Dactylis glanerata Arrhenatherum elatius
Stage 3 Soil pH 5.56	Crataegus monogyna Sambucus nigra Agrostis tenuis - a Artemisia absinthium Arrhenatherum elatius Centaurea scabiosa Chamaenerion angustifolium - a Cirsium arvense - a Dactylis glomerata Equisetum arvense Hieracium sp. Hieracium pilosella - a	Holcus mollis Hypochaeris radicata Linaria vulgaris Lolium perenne Lotus corniculatus - a Luzula campestris Plantago lanceolata - a Poa pratensis Reseda luteola Senecio squalidus Trifolium arvense - a Tussilago farfara Vicia sepium

a = abundant

characterized by the downslope convergence of flow lines tend to be moister and support a more rich and dense vegetation than slopes which are convex in plan and which have divergent flow lines.

The ecology of colliery spoil mounds is also affected by micro-climatic factors. Exposure tends to increase upslope. South facing slopes tend to be several degrees warmer than north facing slopes. Thus, the soils on the higher southwestern sides of pit heaps tend to be markedly drier than those of the northeastern slopes. The high soil-water tensions which are thus built up are also influenced by angle of slope and local wind velocity and cause the wilting of seedlings, thereby restricting plant colonization to lower northeastern slopes (Richardson and Greenwood - 1967).

Pit heaps tend to be composed of grey or black shales which absorb solar radiation. High ground temperatures can be a problem for plants. Thompson and Hutnik (1971) recognize a critical high of 44°C. This high was repeatedly exceeded during their measurements of temperatures on Pennsylvanian spoil banks. These high temperatures also tended to penetrate into the spoil, especially when this was dry (Geiger - 1965). Temperatures of up to 55°C were

TABLE 9.4

VEGETATION OF COAL MEASURE SPOIL:
PLATEAU SPOIL TIPS IN THE BLACK COUNTRY
(after Rees - 1955)

Sides	
Acer pseudo-platanus	Dactylis glomerata
Batula pubescens	Digitalis purpurea - a
Fraxinus excelsior	Equisetum arvense
Quercus robur	Hieracium sp. - a
Salix caprea	Hieracium pilosella
Sambucus nigra	Lotus corniculatus
Crataegus monogyna	Plantago lanceolata
Rubus fruticosus agg.	Pteridium aquilinum
Ulex europaeus	Rumex acetosella - a
	Solanum dulcamara
Agrostis tenuis - a	Senecio squalidus
Anthoxanthum odoratum	Taraxacum officinale - a
Bellis perennis	Trifolium repens - a
Cirsium arvensis	Tussilago farfara
Summit	
Pteridium aquilinum	Digitalis purpurea - a
Chemaenerion angustifolium	Rumex acetosella
Deschampsia flexnosa - a	

a = abundant

TABLE 9.5

VEGETATION OF MacCLANE SHUTE SPOIL TIPS
IN THE BLACK COUNTRY
(Rees - 1955)

Herbaceous Species List	
Agrostis canina	Echium vulgare
Agrostis tenuis - a	Hieracium sp. - a
Arrhenatherum elatius	Holcus lanatus
Chamaenerion angustifolium	Holcus mollis
Cirsium arvense - a	Sagina procumbens
Dactylis glomerata	Tussilago farfara

a = abundant. Soil pH 4.5. Angle of slope 30°.

recorded at depths of 100 mm. Richardson (1958a) has recorded a diurnal temperature variation of 40°C on a bare spoil surface (Table 9.6).

TABLE 9.6

GROUND TEMPERATURE VARIATIONS ON AN UNVEGETATED DURHAM SPOIL TIP
(Richardson - 1958a)

Depth (mms)	Amplitude (°C)	
0	40	(Range 8-48)
762	22	
1524	12.5	
2286	5	

The occurrence of such high surface temperatures also tends to lead to the salinization of the soil surface. However, the extreme temperature regime can be ameliorated by a sufficient development of vegetation (Table 9.7). Sometimes the first stage in this development is the appearance of a moss/lichen community which achieves a sufficient amelioration of surface temperatures to permit invasion by angiosperm primary colonizers.

TABLE 9.7

JULY TEMPERATURES IN VEGETATED AND UNVEGETATED DURHAM SPOIL TIPS
(after Richardson - 1958a)

Cover	Depth (mms)	Amplitude (°C)	Range (°C)
None	0	42	8 - 50
Agrostis sp.	0	12.5	13.5 - 26
Holcus sp.	0	9	11 - 20
None	762	23	10.5 - 33.5
Agrostis sp.	762	4.5	14.5 - 19
Holcus sp.	762	1.25	14.75-16

The ecology of foreign spoil mounds is a process which involves similar principles but different species and environments. Recent studies include those by Cornwell (1971), Schramm (1966), Thompson and Hutnik (1971), Wali (1973) and Schulter (1973).

CHAPTER 10

THE MORPHOMETRY OF NINETEENTH CENTURY FAN-RIDGE SPOIL MOUNDS AND THE EVOLUTION OF RECLAIMED LANDSCAPES

Landforms derived from colliery spoil materials are becoming increasingly common. The future evolution of modern landforms created by land reclamation works may be similar to that already experienced by similar landforms created from similar materials in the nineteenth century. This suggestion is not particularly unusual. In the United States, Smith *et alia* (1966) have illustrated how studies of soil development on old iron ore spoil may be used as a guide to the development of soil on strip-mined spoil banks. Further, the British Association are actively involved in studies which seek to explain the evolution of ancient earthworks in terms of the development of a modern reconstruction--the experimental earthwork at Overton Down (Jewell - 1963; Crabtree - 1972).

The tips considered in this study are located in the Garn Road area of Blaenavon, Gwent. Coal mining began in this area in 1782 and all the tips studied were in existence by 1880. All conform to the fan-ridge type. Fan-ridge coal tips are low mounds which were created by tipping from tramways located on the tip's surface. These tramways were shifted periodically as the tipping front became too far removed from the mine entrance and the result was a spoil tip complex with a characteristic fan-like plan (Figure 10.1, and Jones *et al*. - 1972).

Seventy-two meter-unit slope profiles were measured on fan-ridge spoil mounds in the Garn Road area (Photograph P.10.1). An attempt was made to sample representative slopes on all the main spoil mound complexes in the area and to sample slopes in all available aspect classes. Additionally, measurements were made both of some temporary slopes created by tipping from dump trucks during land reclamation works, and some permanent slopes created by tipping, compaction, and grading for the Dragon Site reclamation project

Figure 10.2 displays the aggregated frequency of meter-unit slope angles greater than 10° recorded on the nineteenth century spoil mounds. The modal class of the histogram is 32°-34°. This class includes those angles usually quoted as the angle of internal friction for dumped coal-spoil (Thompson and Rodin - 1972; Jones et al. - 1972). Steeper angles are not uncommon but they are usually associated with the evolution of terracettes, sheep tracks, and sheep burrows. The histogram is negatively skew.

Two indices have been calculated to express the declivity of the maximum slope segment. These are the three-meter and the five-meter maximum slope angles. Simply, these are the highest mean slope angles recorded from any three or five contiguous meter-unit slope records on any given slope profile. The mean of all the three-meter maximum slope angles calculated was 33.5°. The mean five-meter maximum slope angle was 33.0°. The similarity of these two indices is suggestive of the rectilinear tendencies in fan-ridge spoil mound profiles.

Figure 10.3 illustrates the variation of three-meter and five-meter maximum slope angles. Both data populations are normal. This fact tends to confirm the suggestion that these tips are similar in materials and construction and that, in fact, they constitute a single population. The mean deviation of this population is 3.2° for the three-meter maximum, and 2.8° for the 5-meter maximum slope angle population. The standard deviations are 3.95° for the three-meter maximum, and 3.56° for the five-meter maximum slope angle population.

Chwastek (1970) has demonstrated that it is possible to relate the general declivity of certain Polish lignite mine dumps to their height. However, it proved impossible to discover any relationship between maximum slope angle and mine dump relief.

Recent studies of rates of erosion on slopes of different aspect (Crabtree - 1972; Haigh - 1974a, b) have proved that southwest facing slopes tend to suffer a greater degree of erosion than northeast facing slopes. The variations of both three-meter maximum and five-meter maximum slope angle with aspect is graphed as Figure 10.4. It was not possible to demonstrate any significant relationship. This suggests that on coal tips over 100 years old the main control of maximum slope angle is some function of the angle of internal friction of the spoil materials (Thompson and Rodin - 1972).

Meter-unit slope profiles have also been recorded on a modern spoil mound and on a contemporary reclaimed landsurface. It is interesting to compare these measurements with those from the nineteenth century fan-ridge spoil

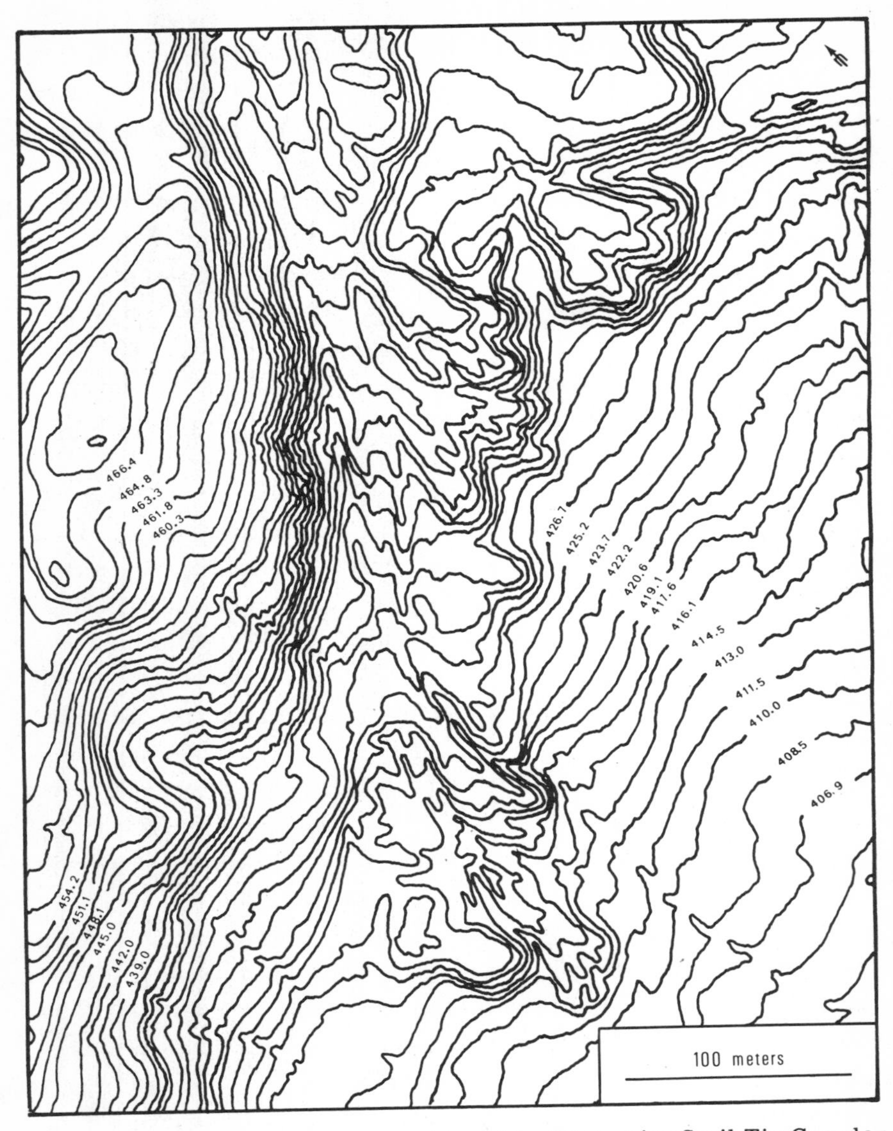

Fig. 10.1.--The Garn-yr-erw Fan-ridge Deep-mine Spoil Tip Complex Study Area.

P. 10. 1. --Fan-ridge Spoil Tips in the Garn Road Area

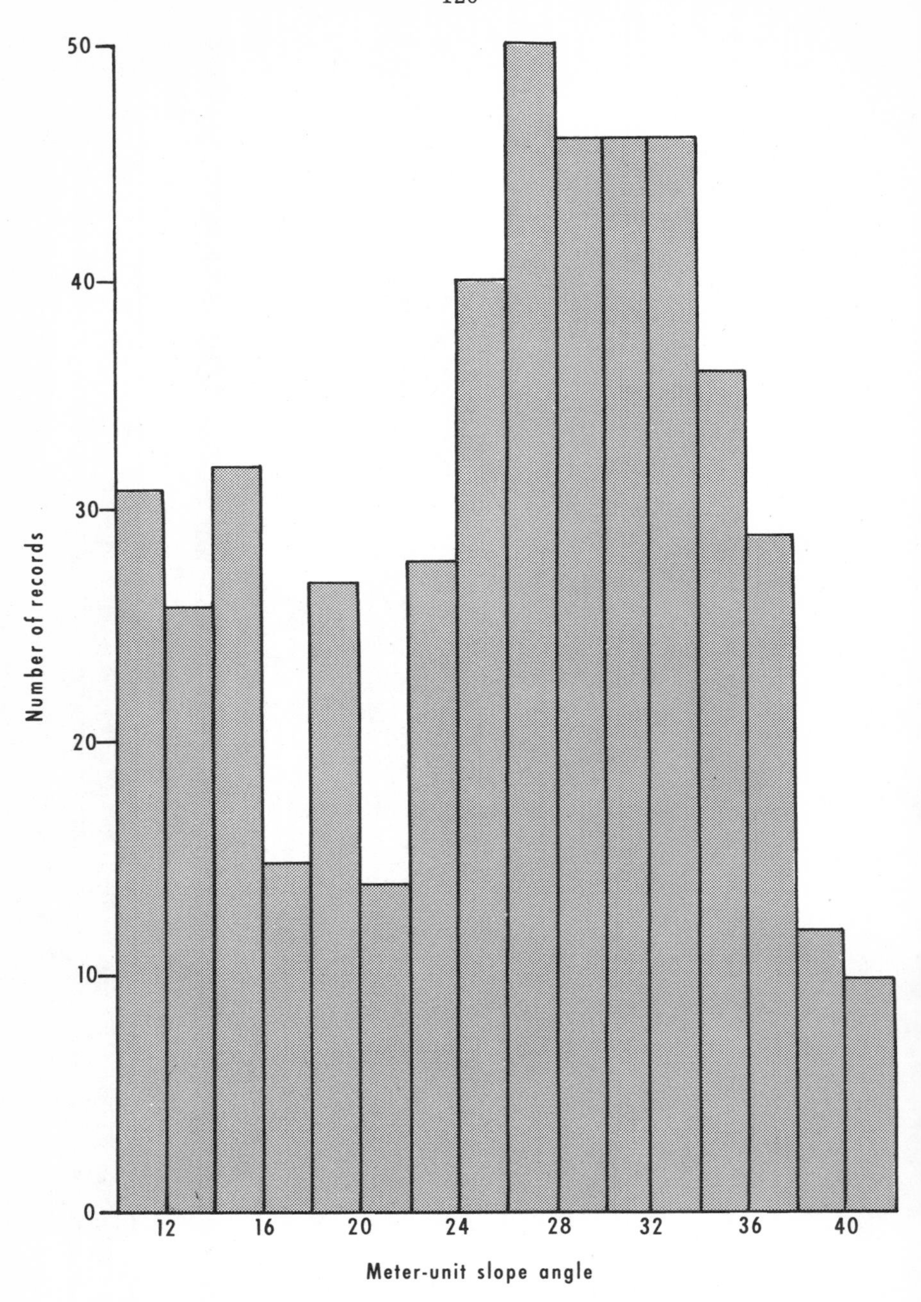

Fig. 10.2.--The Frequency of Slopes over 10°--Garn-yr-erw tips

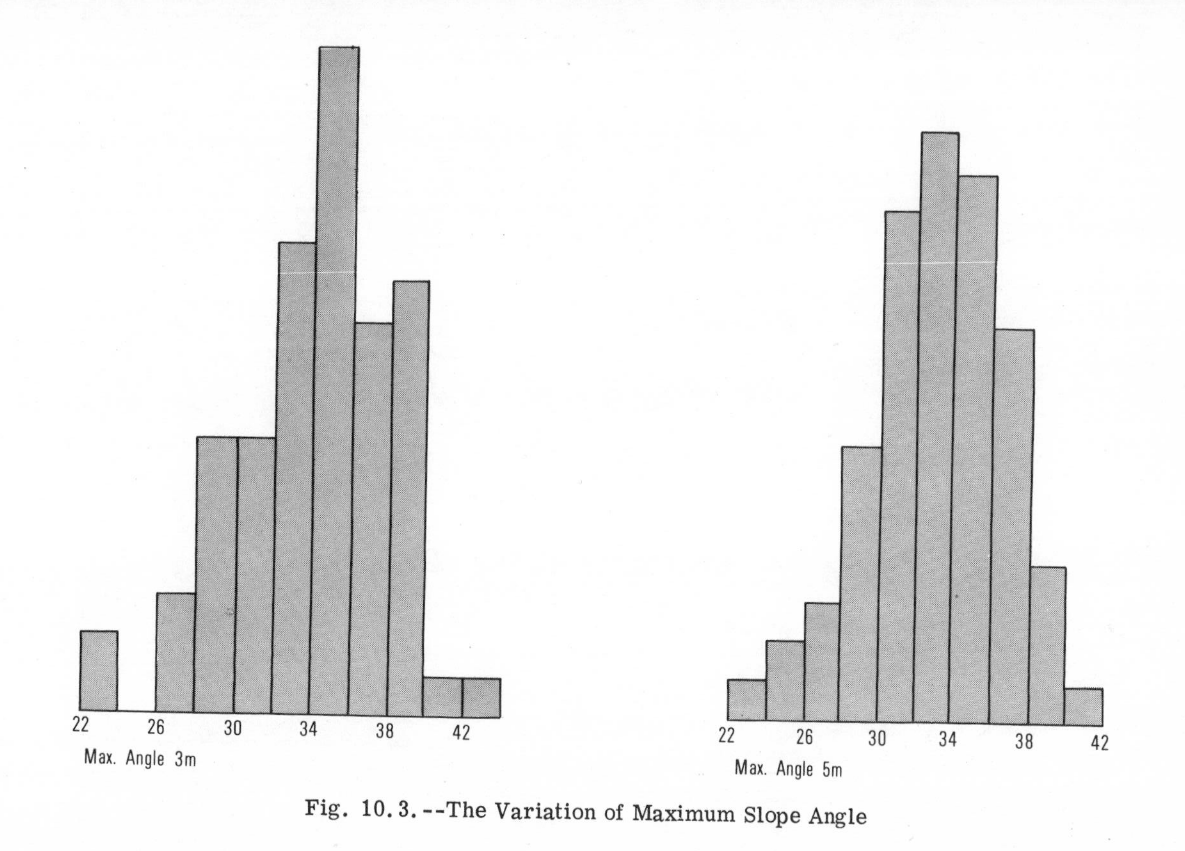

Fig. 10.3.--The Variation of Maximum Slope Angle

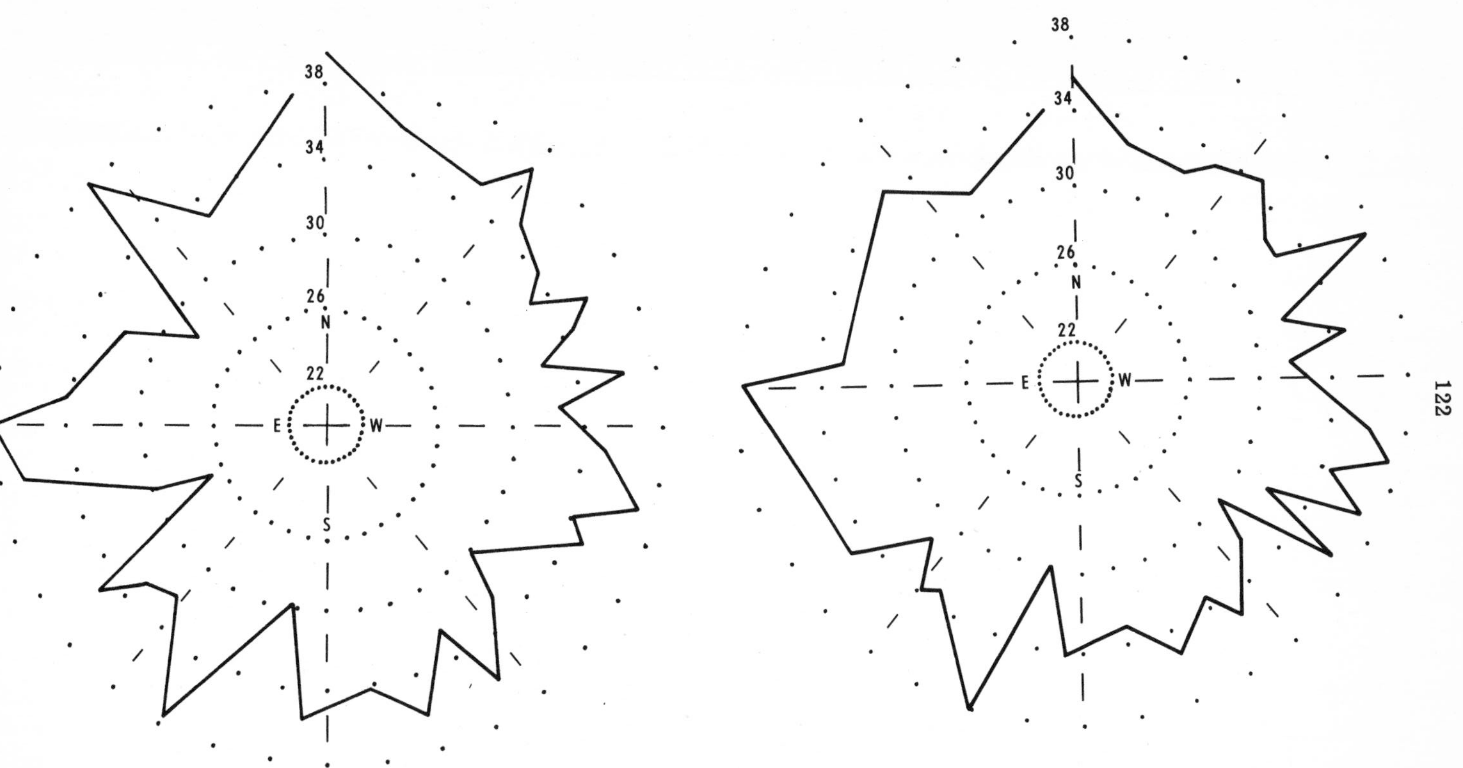

Fig. 10.4.--The Variation of Maximum Slope Angle with Aspect

mounds. Both of the modern slopes were created by a dual process of tipping and compaction which should correspond closely to the technique employed in the construction of the fan-ridge mounds. The modern spoil mound slope probably corresponds most closely to the original form of the fan-ridge dumps. The slopes of the reclaimed surface, which have been modified by grading, are more characteristic of the coal spoil slopes of the future. The average maximum slope angles recorded on the modern spoil tip were 42.2° (3 meter) and 40.3° (5 meter). Those for the reclamation project's slopes were 24.8° (3 meter) and 23.7° (5 meter). None of these results lie beyond the range of fan-ridge tip maximum slope angle variation, but both lie at opposite extremes of that range. The distances of these results from the mean maximum slope angles calculated for the fan-ridge spoil tips, expressed in standard deviation units, were, for the modern spoil mound, +2.2 (3 meter) and +1.9 (5 meter), and for the reclamation project's slopes, -2.2 (3 meter) and -2.5 (5 meter).

Samples of the slope profiles recorded on the modern spoil mound and the Dragon Site land reclamation (Photograph P.10.2) are displayed as Figure 10.5. It can be seen that, in spite of the great differences in actual declivity and minor differences in method of construction, the two types of slope have a certain morphological similarity. The main slopes of both tend to be rectilinear but these straight slopes are broken by small scale steps. Both slopes tend to have well developed basal concavities (cf. Statham - 1973) but the upper convexity is an abrupt junction.

Samples of the slope profiles recorded on the neighboring Garn-yr-erw fan-ridge coal spoil tips are displayed in Figure 10.6. These slopes, again, have rectilinear main slope segments which are sometimes divided by step-like features. Presumably, these features mark junctions between two phases of tipping activity. The slopes have a reasonably well developed basal concavity, but this is dwarfed in all cases by a very well developed upper convexity.

Today, most of the nineteenth century fan-ridge spoil mounds are vegetated. The quality and degree of this plant cover can be related to present spoil tip morphology and to the pattern of slope evolution.

The vegetation of the Garn-yr-erw spoil mounds is quite distinct from that of the surrounding moorland. Usually, it is fairly thin except around the slope foot and, to a lesser extent, on the flat top surface. The vegetation at the slope foot is suggestive of an abundant supply of water. Dominated by *Molinia caerulea* (L.) Moench and *Juncus* spp., it is characterized by a large species list and a thick moss underlayer. By contrast, the tips' summits are covered

P. 10. 2. --Modern Spoil Slopes of the Dragon Site Reclamation

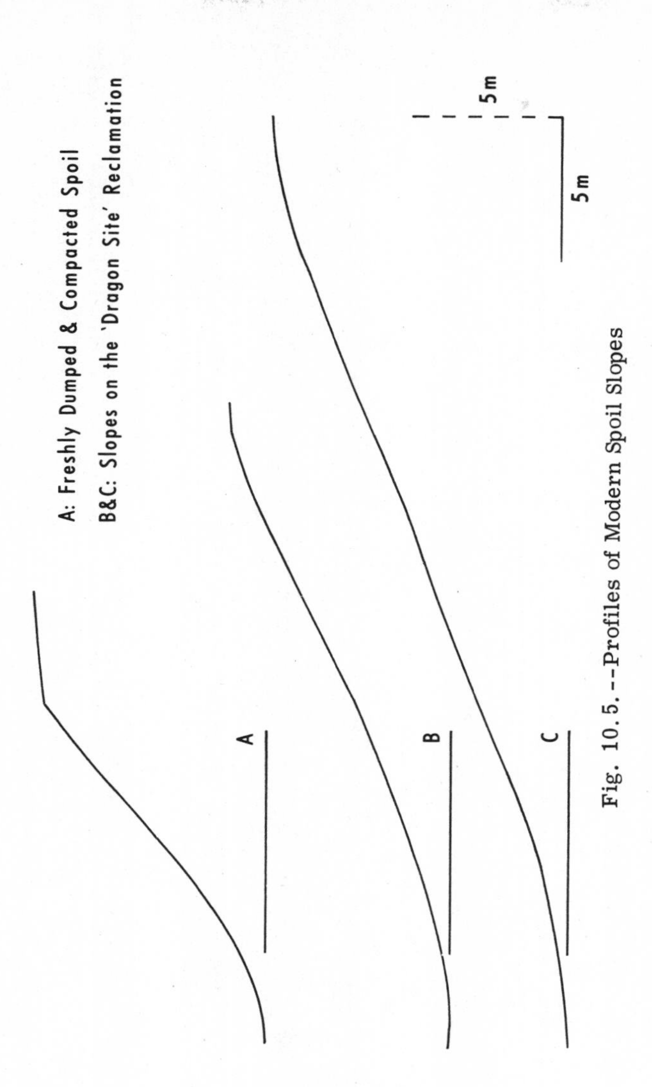

Fig. 10.5.--Profiles of Modern Spoil Slopes

Fig. 10.6.--Meter-unit Profiles, Garn-yr-erw Tips Area (East)

Fig. 10.6.--Meter-unit Profiles, Garn-yr-erw Pits Area (West)

by a thin, dry turf dominated by Festuca ovina agg, Deschampsia flexuosa (L.) Trin., and Agrostis tenuis Sibth. The occurrence of clumps of Nardus stricta L., Calluna vulgaris (L.) Hull, and Vaccinium myrtillus L. is not unusual, and sometimes one or other of these species may be locally dominant. Lichens may form a dense mat amongst the grasses of the tip summit and upper slopes. The most prominent species are Cladonia uncialis and C. arbuscula.

Relatively little vegetation grows on the spoil tip slopes. Frequently, surface cover here does not exceed 50%. The dominant species are Festuca ovina and Deschampsia flexuosa supported by some mosses and lichens. Figure 10.7 illustrates the variations of vegetation cover and species content discovered on two slopes of the Garn-yr-erw tips. These two slopes are morphologically very similar and their vegetation is typical of that found on many of the tips in the Garn-yr-erw area. It can be seen that there is an abrupt decline in angiosperm cover and species content across the lower slope element. Angiosperm plant cover remains uniformly thin across the main slope and upper slope elements. The percentage of bare ground, however, tends to correspond closely with the slope angle. Its decline on the upper convexity is due to the appearance of a dense mat of lichens on the upper main slope. The lichen cover tends to decline upslope across the upper slope elements and reaches a minimum at the junction of the upper convexity with the crest segment.

The profiles examined in Figure 10.7 are both examples of spoil lobe front sites. They are convex in plan and have divergent flow line slopes. Devegetation of the upper convexity is a very common feature of such slopes. An examination of 257 spoil lobes disclosed the fact that while 42% of the tips had a complete plant cover, 34% had developed bare scars on the upper convexity. On 17% of the slopes, this bare scar had extended sufficiently far to constitute a complete devegetation of the spoil tip lobe front and, sometimes, this devegetation extended to affect the arms and even the embayments of the spoil mound complex. The development of scars on other parts of the slope, and in the absence of the development of a bare scar on the associated lobe front upper convexity, was a relatively rare occurrence. Five percent of the slopes had developed scars on the lower concavity of the lobe front but these scars were always associated with either sheep burrowing (Thomas - 1965), or small land slips. Only 2% of the slopes had developed bare scars on the upper convexity of the spoil lobe arm alone. The devegetation of embayment slopes, which possess a downslope convergence of slope flow lines, was uncommon, unless the whole spoil mound was bare. The devegetation of the embayment alone was a

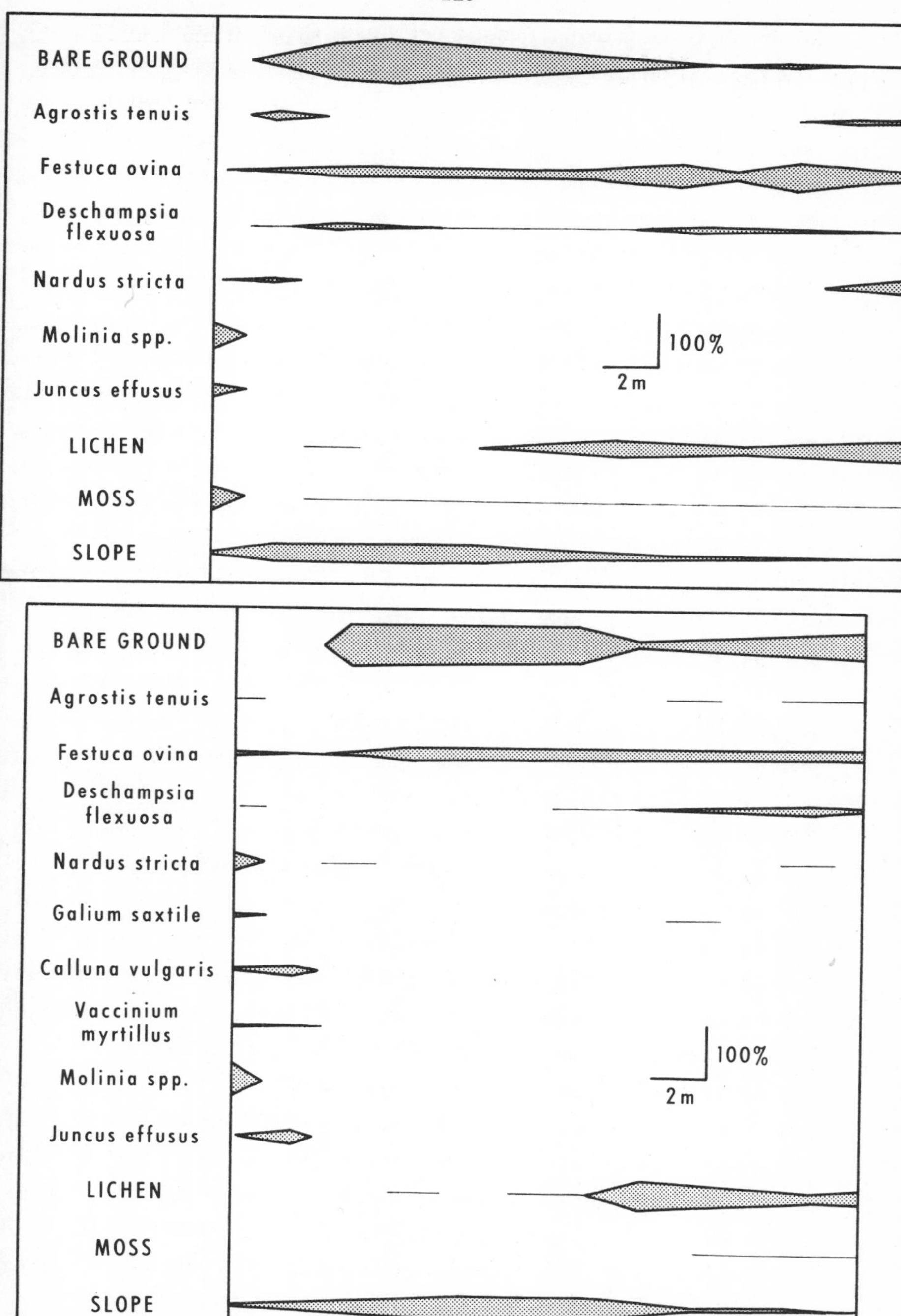

Fig. 10.7.--Distribution of Vegetation on Garn-yr-erw Tip Slope Profiles

very rare occurrence and always associated with the development of a gully channel. However, very few of these spoil tips possess autochthonous gully channels.

These results all suggest that the upper convexity of these fan-ridge spoil lobes is a site of increased erosional activity. This view is supported by a particle size analysis of the surface spoil at different points on the slope profile. Figure 10.8 displays the particle size distributions of three samples from the second slope displayed in Figure 10.7. This particular tip does not possess a bare scar nor is there any marked decline in the vegetation cover at the upper convexity. However, the spoil sample gathered on the upper convexity is very different from two gathered from the mid-slope and lower concavity. The sample from the upper element contains a much smaller proportion of fine debris than either of the other two samples. Since fine particles are more easily removed by erosion than those which are more coarse, it might be suggested that erosion has been, or is, particularly active on this upper slope element.

It is suggested that the development of the bare scar on the upper convexity of these spoil tips is a consequence of the original tip morphology. This was probably characterized by an abrupt junction between the main slope and the tip's flat upper surface (vide: Figure 10.5). This abrupt junction separate the unstable and erosionally active tip face from the relatively stable tip's upper surface. The morphogenetic importance of this junction was enhanced as the tip became colonized by vegetation. The older and more stable upper surface of the tip was capable of supporting a much richer and denser vegetation than the relatively young and mobile surfaces of the upper slopes.

Now, it has been demonstrated that there is an inverse relationship between the speed of erosion and ground retreat and the amount of vegetation (Dragoun - 1969; Holy and Vrana - 1970; Sabata - 1970). It has also been proved that steep slopes tend to suffer a greater erosion than less steep slopes (Zingg 1940; Smith and Wischmeier-1957; but cf. Martin and Rowlinson - 1971). Thus, a consequence of this discontinuity in erosional regime has been a tendency for the development of a free face beneath the stable turf of the tips' upper surfaces. Since this erosional discontinuity is of course greatest on slopes with a down-slope divergence of flow lines (i.e., convex plans), where there is a greater capacity for the downslope transport of surface materials, there has been a tendency for these free faces to develop most frequently on spoil lobe fronts (Haigh - 1974b).

On the Garn-yr-erw tips, free faces have a local relief of some 0.1 or

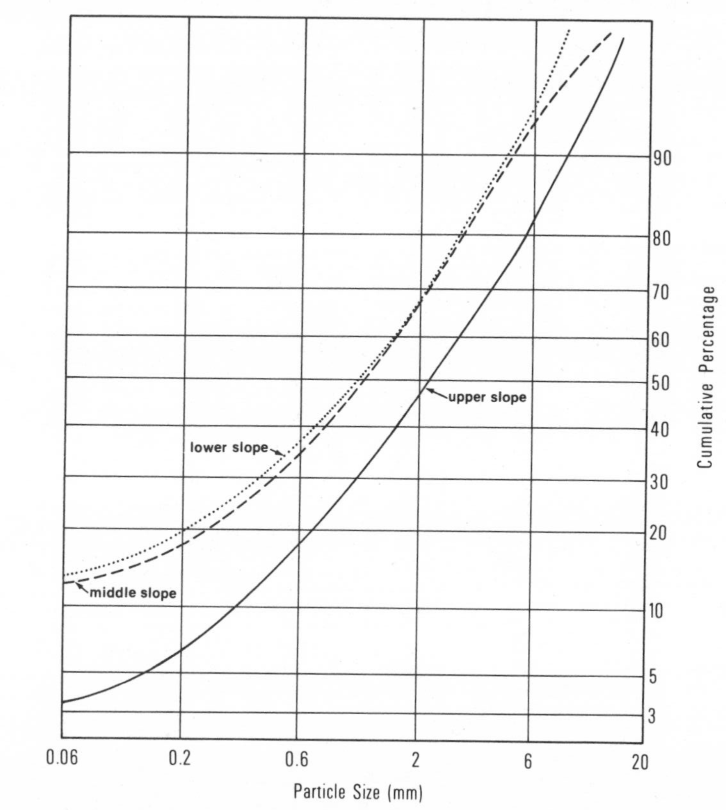

Fig. 10. 8. --Particle Size Distribution, Surface Layers, Garn-yr-erw Tips (cumulative percentage vs. particle size in millimeters).

0.2 meters. However, these heights may be exceeded if a free face is subsequently adopted as a sheep hollow. Free faces affected by sheep burrowing may achieve heights of 0.5 meters.

Once created, free faces have the capacity to extend themselves upslope by the erosion of the spoil beneath the root zone of the tip surface turf, and downslope by burying turf beneath loose fragments. These processes are both accelerated by sheep trampling and burrowing.

To conclude: the existence of an abrupt junction between the densely vegetated flat tip upper surface and the poorly vegetated steep slopes of the coal spoil tip face constitutes a major erosional discontinuity. This discontinuity is thus a zone of great morphological alteration. There has been a tendency for this original abrupt junction to be converted into a broad rounded upper convexity which may affect 40% of the slope profile. Sometimes, erosion at the upper convexity has become sufficiently violent to cause the appearance of a break in the vegetation. This break in the vegetation may extend to affect a large section or even the whole spoil lobe. The appearance of bare scars and free faces is most common on slopes whose flow lines diverge downslope and their development may be exacerbated by sheep trampling and burrowing.

Finally, it may be suggested that similar processes may affect the new slopes currently being created by land reclamation works in this area. These slopes are characterized by an abrupt junction at the upper convexity. It is to be expected that these slopes will evolve by main slope replacement due to the extension of the upper convexity and the amelioration of the upper slope.

A failure of the vegetation on these slopes, should it occur, will commence at the upper convexity on slopes of convex plan. It will be preceded by a decline in the percentage cover of vegetation on the steeper slopes. The risk of devegetation might be reduced if such slopes were, in future, designed with well rounded upper convexities which would be less liable to morphological alteration (cf. Young and Mutchler - 1968, 1969; Meyer and Kramer - 1968, 1969).

CHAPTER 11

GROUND RETREAT AND SLOPE DEVELOPMENT ON THE PLATEAU SPOIL TIPS AT MILFRAEN

The plateau spoil tip complex at Milfraen lies at 455 meters on open moorland at the foot of the Pennant Sandstone Scarp of Coity Mountain (N. G. Ref: 218101) (vide: Photograph P. 11. 1). Milfraen's coal-workings date from 1846. However, the visible surfaces are much more recent and probably date from the 1920s. The colliery itself was effectively shut down by a mining disaster in July 1929 (Photograph P. 11. 2) though it did not finally close until 1931. Today, the 17-acre Milfraen site is totally abandoned and even the old miners' cottages there have been destroyed. The area was scheduled for reclamation as rough grazing by the Monmouthshire Derelict Land Reclamation Joint Committee's "Comprehensive Report" of 1971.

Morphologically, the plateau spoil tips at Milfraen are fairly typical of their type. Their local relief ranges from 10 meters to 25 meters. Slopes are steep and tend to be rectilinear. They attain maximum angles of 32-38° and these may be maintained over 50% of the slope profile. Characteristically, basal concavities are abrupt. The transition from main slope to moorland rarely requires even 20% of the slope profile. The transition from main slope to spoil tip plateau surface is far more gradual. Frequently, the upper convexity comprises 30% and 40% of the profile's length. Total slope length ranges from 20 meters on south facing slopes to 50 meters on slopes which face north. The tip plateau, which appears perfectly flat from a distance, in fact, possesses a local relief of 2-3 meters as a result of subsidiary tipping and the location of past tramways.

The coal spoil at Milfraen is fairly fine and generally homogeneous. It may contain, locally, a small admixture of very fine washeries waste from the small Milfraen washery, though all the evidence suggests that the bulk of this was dumped separately. The particle size rating for five spoil samples taken from different parts of the Milfraen complex is shown as Figure 11. 1. One lobe of the Milfraen complex has been burnt. However, this area was not included by these studies.

P. 11.1.--The Plateau Spoil Tip Complex at Milfraen

P. 11. 2. --Milfraen Colliery, July 1929: The day of the mining disaster

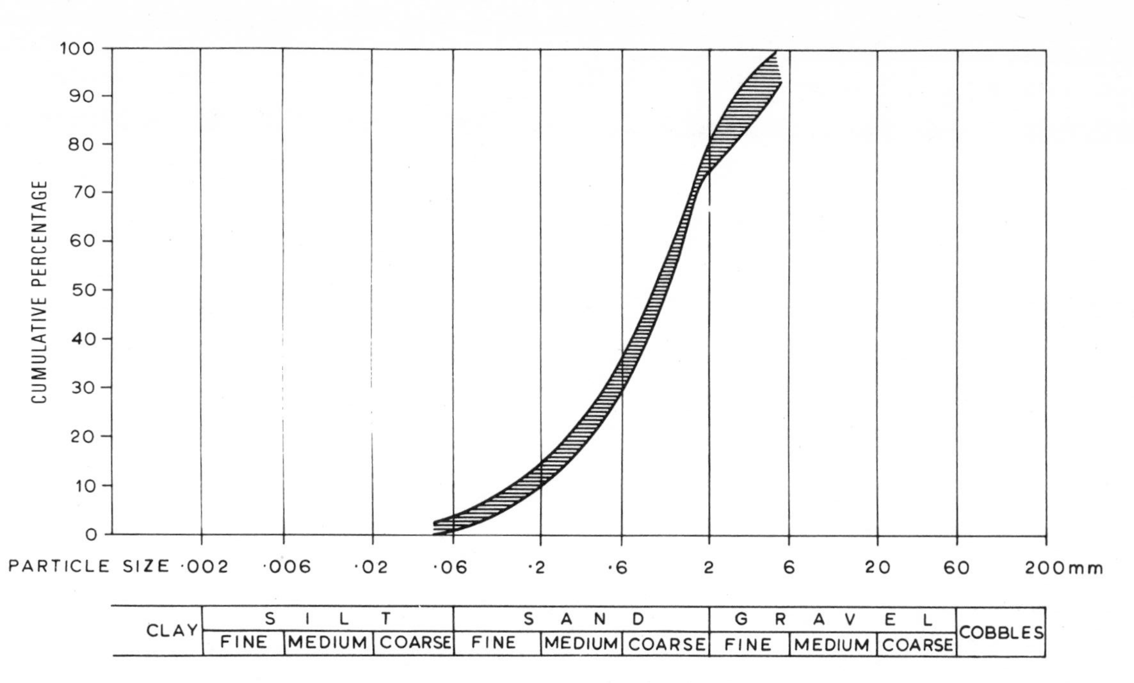

Fig. 11.1.--Milfraen--Particle size distribution of surface spoil (range of 5 samples)

Most of the Milfraen tip complex is vegetated and covered by a close sheep cropped turf. This turf is quite dense on the tip plateau and at the slope foot (cover: 80% plus). Perhaps a dozen or so species are represented at any one site. These include: Festuca ovina agg, Agrostis tenuis Sibth, Galium saxatile L., Luzula campestris (L.) DC., the moss: Polytrichum, several species of thistle and other ruderals. The heart of the tip plateau also supports clumps of Nardus stricta L., Deschampsia flexuosa (L.) Trin., and Juncus effusus L. The vegetation of the tips' slopes is far thinner (cover: 50% or less) and contains far fewer species. Festuca ovina dominates and this is supplemented by Agrostis tenuis with a little moss and some lichen, notably on the north facing slopes.

It is likely that, in the past, all of the slopes of Milfraen were covered by some sort of vegetation. The denuded slopes of the present day show evidence of past vegetation cover as turf walls at the head of the slope and residual mid-slope patches. It is possible that trampling by sheep and other livestock has greatly contributed to the loss of vegetation from parts of the tip complex. There is a considerable development of terracettes, sheep tracks and sheep hollows (Photograph P.11.3 and cf. Thomas - 1965). The tips are heavily grazed by sheep, cattle, and horses, and also support a small rabbit warren. Photographs P.11.4 show the churning effect of sheep trampling on a fixed meter-square quadrat established on a north facing slope. Photograph P.11.5 illustrates a final stage in the destruction of the turf cover in another part of the same slope. Here, the turf has finally been reduced to tufts. These tufts protect the spoil beneath from erosion by rainsplash and sheet-flood and thus become elevated above the level of the surrounding spoil. The development of these islands interferes with sheet-flood and the migration of surficial materials which tend to accumulate on the upslope side of the grass tuft. This accumulation is balanced by scour on the downslope side of the tuft (Figure 11.2). The result, which is, no doubt, exacerbated by frost-heave and soil creep is that the tuft tips forward and begins to roll (Photograph P.11.6). This rolling proceeds until eventually the tuft is completely overwhelmed by the upslope detrital accumulation and the grass is killed.

Precipitation at Milfraen is heavy and persistent. A nearby W.N.W.D.A. rainguage has been recording an average annual rainfall of about 1430 mms since 1968. The bulk of the rainfall at Milfraen falls in the winter months. The highest recorded monthly totals during the period of direct observation were 313 mms and 284 mms in January and February 1974, and 281 mms in

P. 11. 3. --Milfraen--Sheep hollow and terracettes (with 140 x 230 mm file card scale).

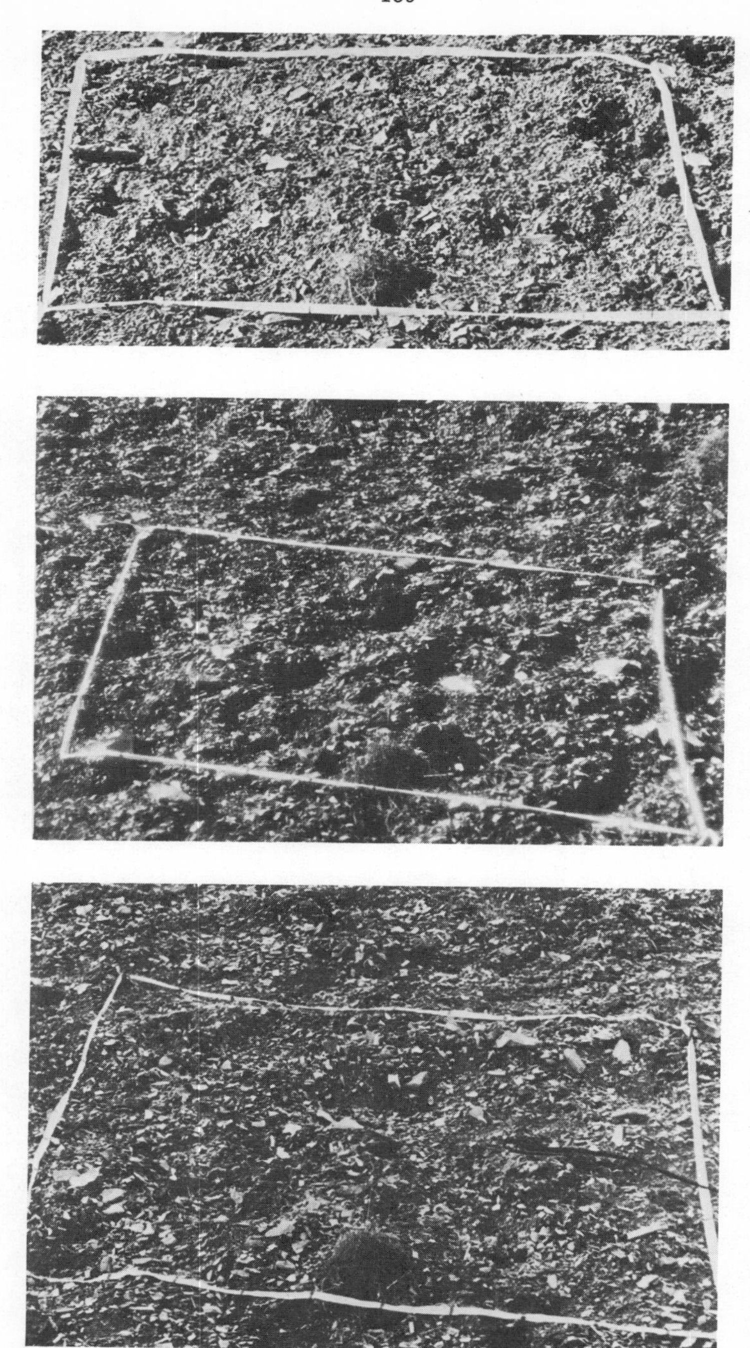

P. 11. 4. --Sequential Photographs of Effects of Sheep Trampling at Milfraen

P. 11. 5. --Grass Tufts on a Devegetating Shale Slope at Milfraen

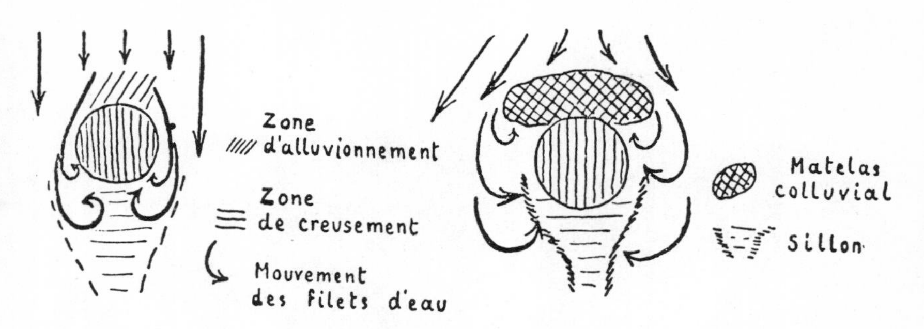

Évolution de l'érosion au pied d'obstacles.
à gauche : Premier stade. *à droite* : Deuxième stade.

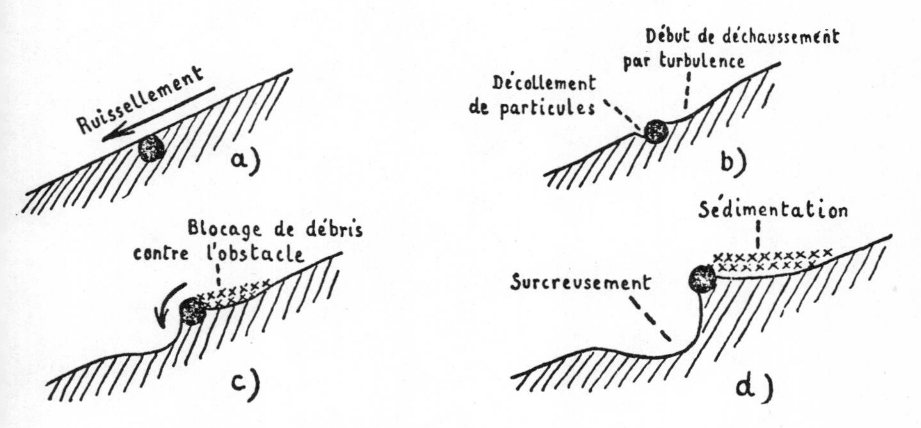

Création, par le ruissellement, des marches au niveau des racines.

Fig. 11.2.--Effects of Surface Obstacles on Sheet Flood Erosion (from: G. Rougerie - 1960).

P. 11. 6. --Surficial Wash and Creep Processes, Inhibited by Grass Tufts Rooted in More Stable Lower Soil Layers, Destroy Such Tufts by Downslope Undermining, Rolling, and Overwhelming beneath Upslope Debris Accumulations.

December 1972. Rainfall recorded during the course of this study is graphed as Figure 11.3.

Ground retreat is being monitored on three slopes at Milfraen. These slopes have been selected in an attempt to examine the effects of aspect, vegetation and gullying on slope development. Two of the profiles support a plant cover. "Milfraen N.E." (Photograph P.11.7) has a northeastern aspect and "Milfraen S.W." (Photograph P.11.8) faces southwest. These profiles were both instrumented in July 1972. The third profile was not established until March 1973 and this is located on an unvegetated slope. "Milfraen E." (Photograph P.11.9) has a neutral aspect and its instrumentation has been extended to span a small gully channel.

Gully channels are not common at Milfraen any more than they are common on the other plateau and "fan-ridge" spoil tips (Jones et al. - 1972) of the Blaenavon area. The whole of the Milfraen spoil tip complex supports only two permanent channels. Both occur in the embayments between smooth unvegetated slopes.

Ephemeral rills developed in all three summers of the experiment. Rills tended to develop after periods of particularly intense rainfall and tended to persist throughout the summer. Sometimes further intensive rainfalls reactivated the rill channels, but there was no evidence for more than two or three phases of rill dissection during any individual summer. Rills formed during the summer months did not persist long. There was a tendency for them to be destroyed by winter frosts as predicted in Schumm's description of the rill cycle (Schumm - 1956a). Photographs P.11.10 illustrate the destruction of rills during the winter of 1972-1973. Rills which develop in subsequent years do not necessarily adopt the same channels as those excavated by the rills of the previous season. Rill development seems to be assisted by high rainfall intensities and low soil moisture content but is totally inhibited by diurnal freeze-thaw.

The instrumentation employed by this study consisted of erosion pins of two different lengths: 610 mm (cf. Schumm - 1956, 1967; Schumm and Lusby - 1963) and 153 mm (cf. Bridges - 1969; Bridges and Harding - 1971; Clayton and Tinker - 1971; Emmett - 1965). Instrumented profiles were laid out as follows: a tape was laid down along the line of steepest slope. Rows of erosion pins were established at right angles to this line. Each row was one meter wide and bounded by two 610 mm rods. Three 153 mm rods were spaced out at 250 mm intervals between. Milfraen N.E. and Milfraen S.W. were both laid out according to this standard plan. Milfraen E. was established as a double

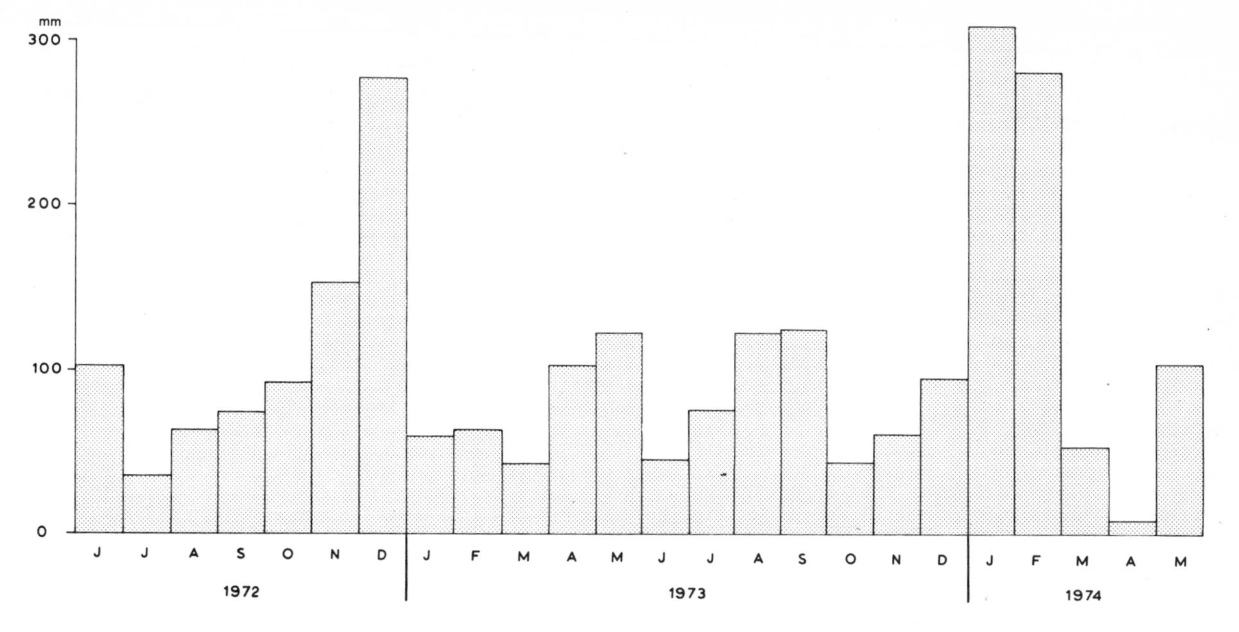

Fig. 11.3.--Rainfall June 1972-May 1974, Llanelly Hill, Waunafon. W.N.W.D.A. raingauge no. 3303. Height: 432 meters. Distance from Milfraen: 800 meters.

P. 11. 7. --Site of Milfraen N. E. Erosion Grid

P. 11. 8. --Site of Milfraen S. W. Erosion Grid

P. 11. 9. --Site of Milfraen E. Erosion Grid

P. 11. 10. --Winter Rill Destruction on Milfraen Spoil Banks

grid with internal spacings expanded by a factor of three. Grid rows were not evenly spaced along the slope profile except on Milfraen E. Elsewhere, erosion grid rows were grouped in areas where the rate of change of slope angle was greatest. All grid row spacings were multiples of a 1.5-meter unit, the typical spacing was 3 meters. Two records were measured from each erosion pin, one to its left and the other to its right. Measurements were recorded to the nearest millimeter. The accuracy of the recording method was calculated from repeated recording experiments. The standard deviation of the repeated record from the original was 0.7 mm and the mean deviation was 0.5 mm per individual erosion pin.

The results of this experiment are probably best introduced by a consideration of the "quarterly ground retreat estimates" for each of the instrumented profiles. The "estimate" is a function designed to represent the average retreat per unit area over the whole slope profile in a standard three month period. It is calculated from the formula:

$$Q = \frac{91}{t}\left(\frac{\Sigma\,[rw]}{\Sigma\,[w]}\right)$$

where: Q is the quarterly ground retreat estimate;

t is the time in days between data records;

r is the mean difference in exposure of the erosion pins in each grid row;

w is a weighting factor representing the distance separating any given grid row from its neighbors;

91 is, of course, the approximate number of days in a quarter of a year.

Cumulative quarterly retreat of the instrumented slopes at Milfraen is presented as Figure 11.4. To enable an easier comparison of the record from each of the two years' data from Milfraen N.E. and Milfraen S.W. with the shorter record from Milfraen E. all results have been set to zero in June 1973. The graph is a record of ground retreat. Hence, a rising trace indicates that the ground surface is becoming lowered and a falling trace indicates that the ground surface is gaining in altitude.

The pattern of generalized ground retreat is very much as one might have predicted. The unvegetated slopes of Milfraen E. suffer a far more rapid ground retreat than do the vegetated slopes of Milfraen N.E. and Milfraen S.W. Some of the slopes of Milfraen E. grid have a basal gully control. This tends to prevent slope foot accumulations of detritus and actively stimulates retreat by undercutting and steepening the slope foot. It is not surprising, therefore,

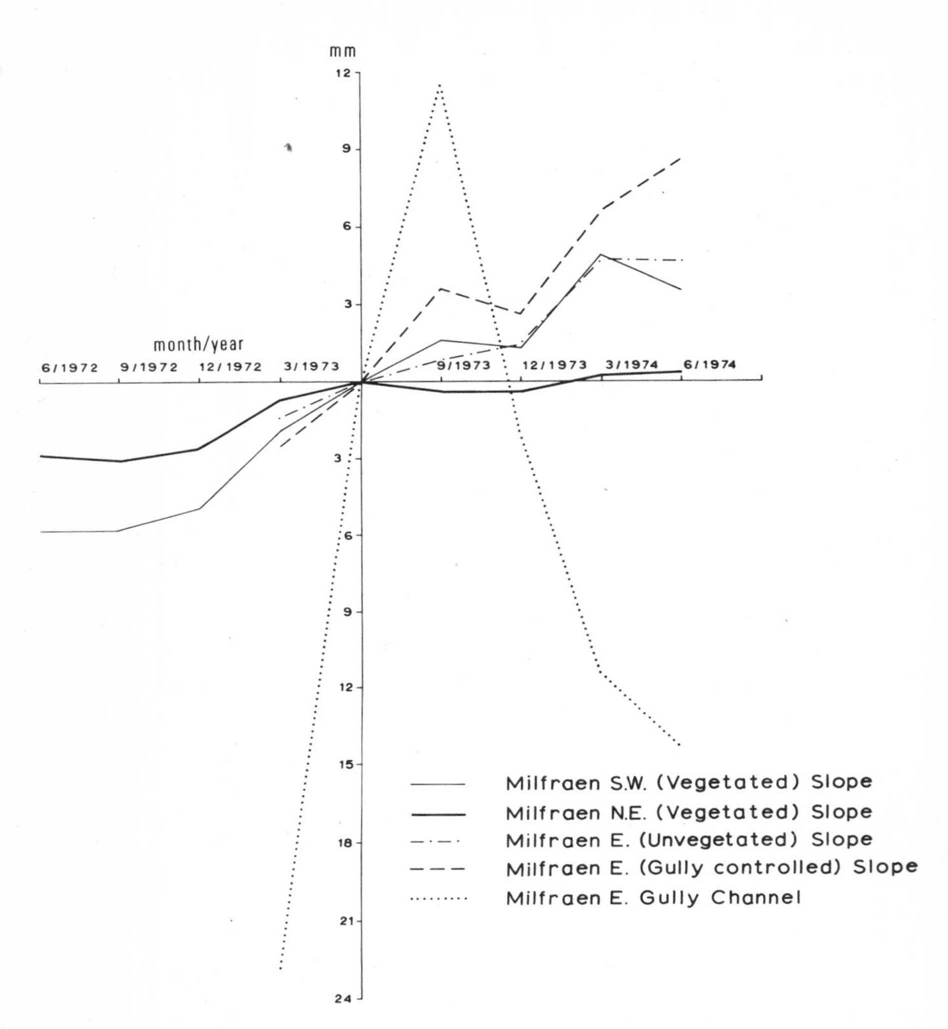

Fig. 11.4.--Milfraen Tips: Mean ground retreat on each erosion grid, 1972-1974. All results are based at zero for June 1973. An upward sloping graph denotes ground retreat.

to discover that it is the gully controlled slopes which suffer the greatest ground retreat. Milfraen E.'s gully channel was the most active of the slope profiles examined, but here erosion was mainly balanced by deposition--presumably, the deposition of spoil removed from the gully controlled slopes. Of the vegetated slopes, Milfraen S.W. retreated faster than Milfraen N.E. This was true in both years of the experiment and the results echo Crabtree's observations of slope decay on the British Association's experimental earthwork at Overton Down, Somerset (Crabtree - 1972; Jewell - 1963).

The "ground retreat estimate" is very much a generalization. Accordingly, two series of graphs have been prepared to show the within-profile conditions of ground retreat. In the first series of graphs, the mean retreat of each individual grid row in the instrumented profile is graphed against time. The locations of each of these individual erosion grid rows and their relationships to the slope profiles under consideration are displayed as Figure 11.5. The graphs of ground retreat per grid row are presented as Figure 11.6, Milfraen S.W.; Figure 11.8, Milfraen N.E. and Figures 11.10, 11.11 and 11.12 for the gully independent or interfluve slopes, the gully controlled slopes, and the gully channel slope, respectively of Milfraen E. In the graphs of the second series, the mean retreat of each individual grid row is graphed against distance upslope for each three monthly recording period. Two graphs are included here: Figure 11.7, Milfraen S.W., and Figure 11.9, Milfraen N.E.

Taken together, the graphs in Figures 11.6-11.12 emphasize three important points. First, any slope section may, at one time or another, experience phases of ground advance, ground retreat, or neutrality, though one of these conditions may be more usual than the others. Second, slope sections experiencing ground retreat, ground advance, and neutrality, characteristically coexist on a single slope at the same time (cf: Pinczes - 1971; Marosi - 1971). Third, the speed and differentiation of advance/retreat activity on the slope profile varies during the year (cf: Schumm - 1964; Schumm and Lusby - 1963). Ground retreat tends to be greater and affect more of the slope during the winter months, December-March. By comparison, the autumn quarter, September-December, is a period of relative inactivity when slight ground advances affect considerable sections of the slope.

By and large, in these experiments, it is possible to relate, with confidence, the larger advance and retreat values to deposition and erosion respectively. Smaller secular changes may owe a lot to soil creep while small seasonal changes may merely be the result of soil moisture changes. The slight

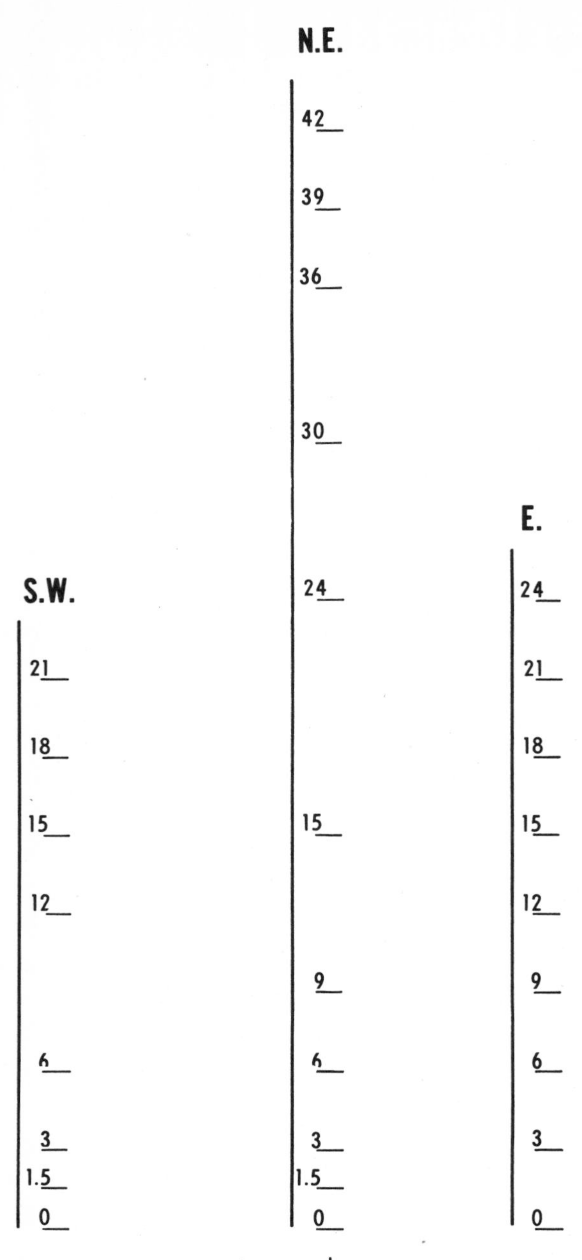

Fig. 11.5.--Erosion Grid Format--Milfraen

ground advance recorded on most profiles in autumn, for example, may be due solely to the fact that the spoil dries out in summer but remains moist throughout the rest of the year. The autumn ground advance may therefore be a result of spoil expansions caused by an increasing soil water content. Schumm and Lusby (1963) discovered from studies in Colorado that their erosion pins tended to be exposed most in fall and least in spring. They related this phenomena to summer rainfalls which tend to compact and seal slope surfaces and winter frosts heaving and loosening the soil surface. In this capacity, needle-ice formation seems especially note-worthy.

Collectively, the graphs from the vegetated profiles, Milfraen N.E. and Milfraen S.W., may be considered to be more representative of the normal pattern of erosion on the plateau spoil tips of the Blaenavon area. Most of these tips are vegetated and most support a similar vegetation. In fact, the results from the two vegetated profiles have a great deal in common.

Slope foot deposition is a characteristic of both slopes in both years. On the smaller and more active slope, Milfraen S.W. (Figs. 11.6 and 11.7), ground advance affects all of the lower slopes and even neutralizes the 6-meter trace in 1972-1973. This effect however, is dramatically absent in 1973-1974 when the 6-meter grid row emerges as that which registers the greatest retreat (14.4 mms). The influence of this erosion is transmitted downslope and the 0-meter, 1.5-meter and 3-meter graphs record a correspondingly greater advance in 1973-1974 than they did in 1972-1973. Figure 11.7 demonstrates that this erosion was mainly effected in the December 1973-March 1974 quarter, while deposition of the eroded material is not apparent until March-June 1974. The lower slopes of Milfraen N.E. (Fig. 11.8) behave, essentially, in a similar fashion. Once again, the main records of ground advance are found in the second year's data. However, there is no evidence that ground advance/neutrality affects a greater proportion of the lower profile in 1972-1973. In fact, in 1972-1973, the trace which registers the greatest retreat on the whole profile is the 1.5-meter graph (9.1 mms), though this does return to pattern in the following year (-3.5 mms). The 3-meter trace, however, records a slight net retreat in both years of the experiment (1972/3: 4.0 mms; 1973/4: 2.7 mms). Thus, on Milfraen N.E., only the 0-meter graph consistently registers ground advance (1972/3: -3.2; 1973/4: -12.4). One reason for the comparative absence of extensive slope foot accumulation on Milfraen N.E. may be the relative inactivity of this profile (cf: Fig. 11.4). Another reason may be that the lower slopes of Milfraen are traversed by a zone of heavy trampling. A band of

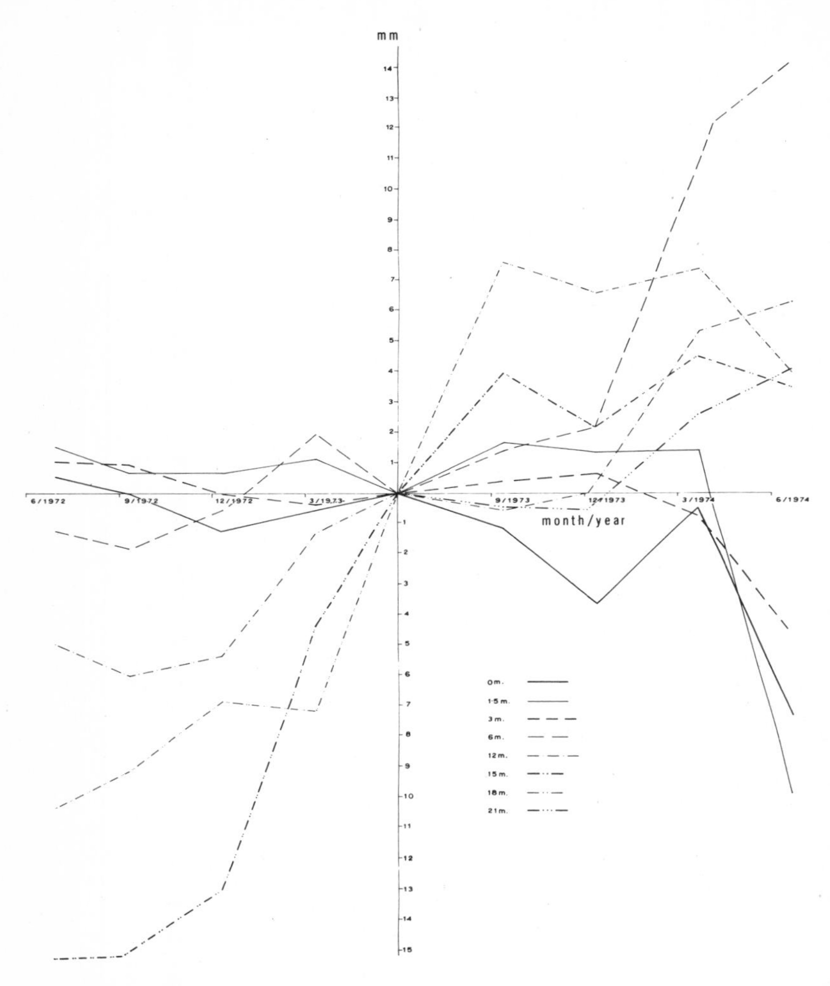

Fig. 11.6.--Milfraen S.W. Grid. Ground retreat per grid row: June 1972-June 1974. All results set to zero on June 1973.

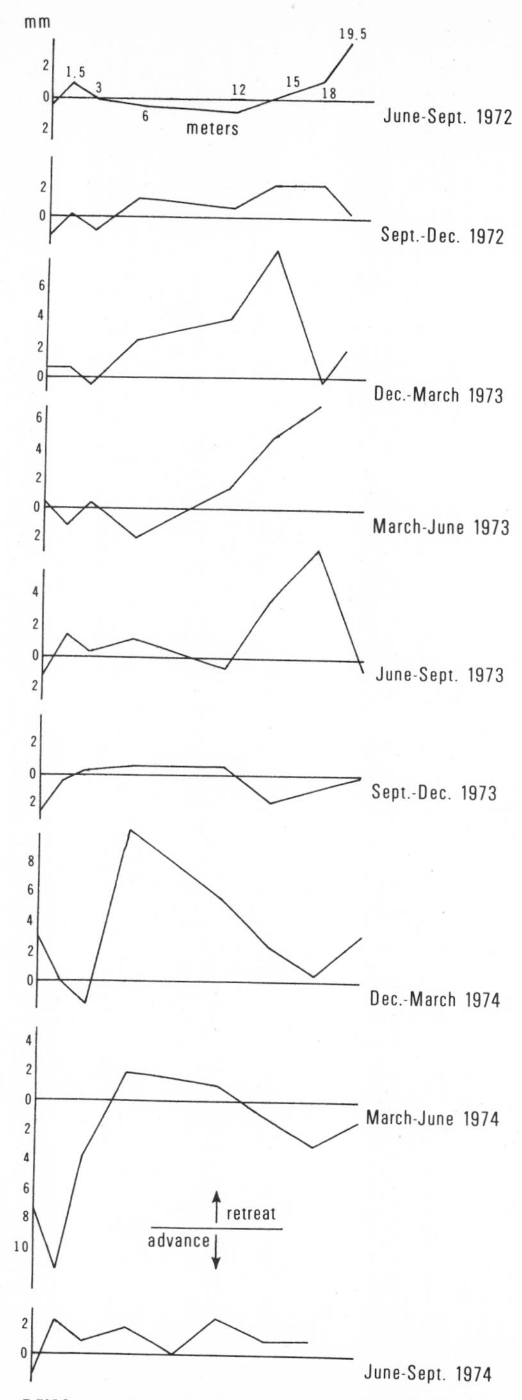

Fig. 11.7.--Milfraen S.W.: Upslope variations of ground retreat recorded during each quarter year of observation. A positive graph depicts ground retreat.

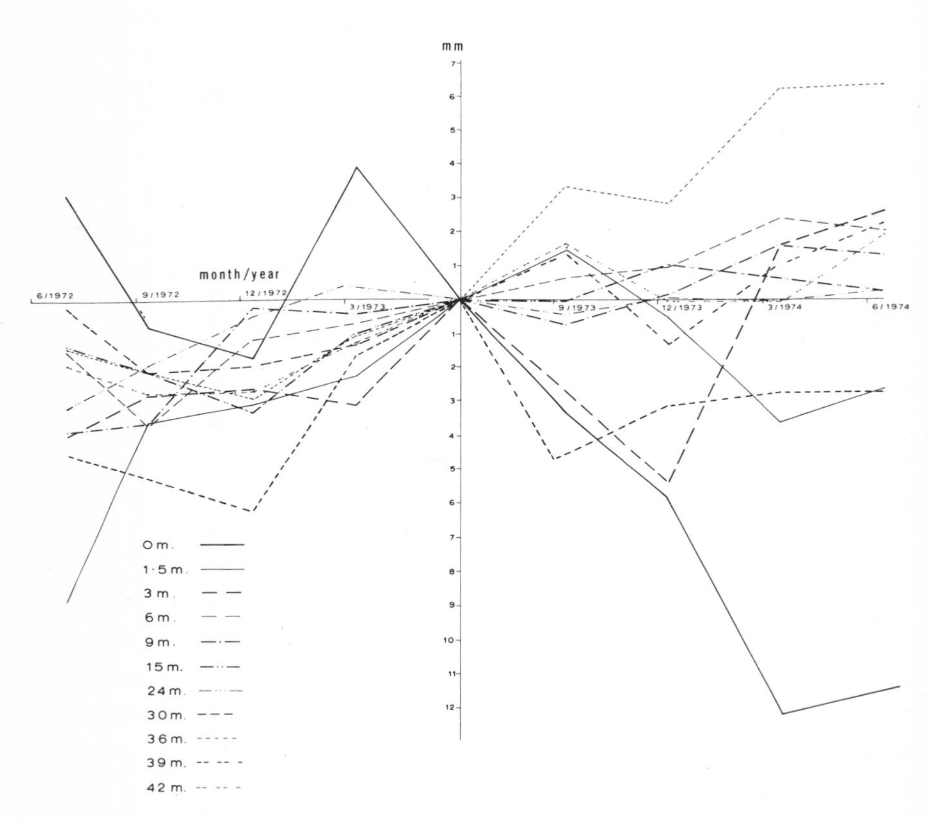

Fig. 11.8.--Milfraen N.E. Grid. Ground retreat per grid row: June 1972-June 1974. All results set to zero on June 1973.

gently sloping and firm ground separating the precipitous spoil tip slopes from an unpleasant bog, it is crossed by two well-used sheep tracks (Photograph P.11.11). Trampling might reduce apparent slope foot ground advance by compaction while scuffing would tend to remove loose debris from the immediate slope foot.

There is a tendency towards parallel retreat on the main slope sections of the two slope profiles. On Figures 11.6 and 11.8, the graphs from main slope grid rows tend to be grouped in the center of the spread of ground retreat traces. Further, the retreat registered by the mid-slope graph tends also to be that which lies nearest to the center of the whole range of retreat scores (Fig. 11.6: 12 meter; Fig. 11.8: 15 meter, 24 meter). Figure 11.9, especially, highlights the tendency for main slope grid rows to yield uniform low positive or neutral records for ground retreat. This tendency is also apparent on Figure 11.7 in the graphs for June-December, 1972, and September-December 1973, but elsewhere on this smaller and more active slope, the effect is masked by interference emanating from the morphogenetic activities of the upper convexity and basal concavity.

High retreat values are recorded at either extreme of the main slope. Sometimes, very high values are recorded at the lower end of the main slope. These seem to have no morphological importance. Frequently, as was the case with the extreme scores from the 6-meter trace of Milfraen S.W., this is a legacy remaining from a previous temporary extension of the slope foot depositional regime occasioned by a particularly massive erosion of the upper convexity. Higher retreat scores are typical of the upper convexities of both slopes. In fact, the upper convexity of Milfraen's plateau spoil mounds seem to be areas which are uniquely vulnerable to erosion and morphometric change. Several of Milfraen's spoil lobes have developed bare scars on the upper convexity of an otherwise well vegetated slope, and again, as was the case with Blaenavon's fan-ridge spoil tips, there is a tendency for the development of bare scars to be associated with slopes of convex plan. It is possible that an initial rupture in the vegetation of the upper convexity, perhaps due to sheep trampling at the edge of the tip plateau, is capable of extending itself: downslope by burying the thin mid-slope turf beneath a welter of loose fragments, upslope by undermining the turf of the plateau margin. A turf-capped free face some tens of millimeters high has developed on the crest of Milfraen N.E. and it seems likely that the devegetation of Milfraen E. developed from just such a turf rupture. Here, the upslope turf wall is still higher and located perhaps half a me-

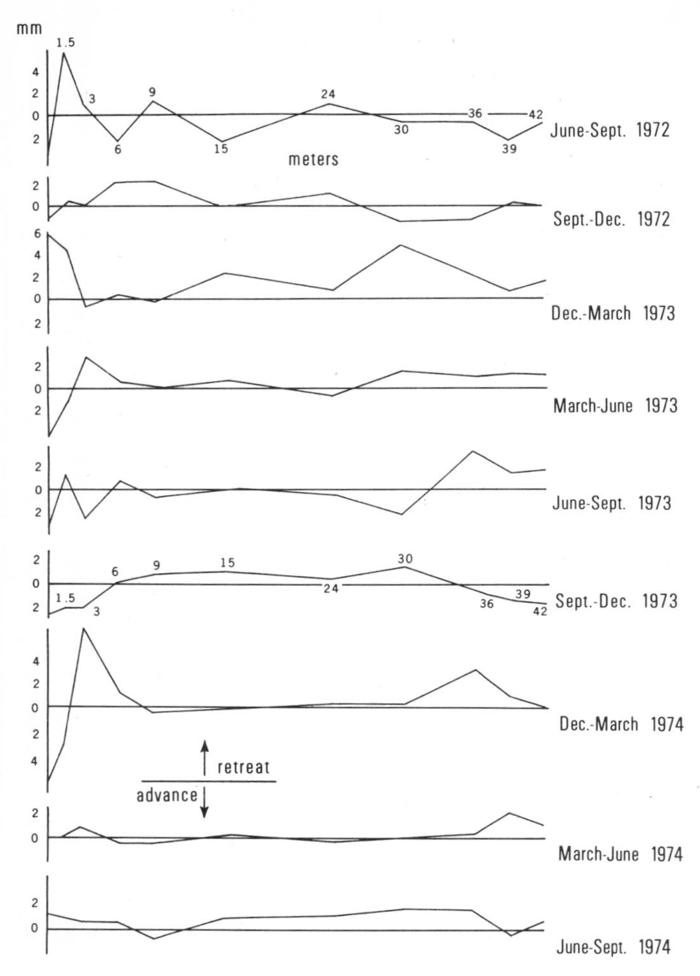

Fig. 11.9.--Milfraen N.E.--Upslope variations of ground retreat recorded during each quarter year of observation. A positive graph denotes ground retreat.

ter from the actual head of the slope. The development of part of this turf wall owes something to the process of "sheep burrowing" (Thomas - 1965), though most of these scars have developed in situations which are too exposed to offer much appeal as sheep shelters.

Results from the unvegetated, gully-independent or interfluve, slopes of Milfraen E. (Figure 11.10) are somewhat different from those recorded on the vegetated slopes. The most important difference between the three sets of graphs (Figures 11.6-11.10) is that there is no evidence that it is the upper convexity of the unvegetated slope which suffers most erosion. On the contrary, the peak retreat scores on the gully-independent slopes of Milfraen E. are all registered by mid-slope sites (Figure 11.10: June 1973-June 1974: 9 m--10.8 mms, 12 m--10.0 mms, 15 m--10.8 mms). The evenness of the results, again, indicates a tendency for parallel retreat mid-slope. Retreat scores for the upper convexity are also high--but not so high (Figure 11.10: June 1973-June 1974: 18 m--7.8 mms, 21 m--10.3 mms). Ground advance affects the lower 3 meters of the profile, and predictably, results from the spoil plateau suggest neutrality or slight retreat.

The term "gully-controlled" is here applied to data gathered from erosion pins situated on slopes whose flow lines terminate in the gully channel. These slopes are immediately adjacent to those interfluve slopes described in the previous paragraph and consequently differences between these two sets of data can only be a result of the activities of the gully channel.

There are several major differences between the patterns of ground retreat recorded on the gully independent slopes (Figure 11.10) and the gully controlled slopes (Figure 11.11). The first is simply the fact that retreat was more active on the gully-controlled slopes. A direct comparison of the retreat scores for each grid row during the period June 1973-June 1974 shows that greater retreat was suffered by the gully-controlled slopes in six of the nine grid rows. The three exceptions were the 6-meter row and two upper slope rows at 21 meters and 24 meters. The differences between the two sets of graphs tend to be larger nearer the gully channel. This is simply because the gully activity has tended to prevent the accumulation of debris at the slope foot.

In fact, the gully may play a more active role. The peak retreat scores recorded on the gully-controlled slopes tend to be 50% higher than those recorded on the gully-independent slopes and they also tend to occur lower down the slope profile. This suggests that the gully may be actively accelerating erosion by undercutting the base of the slope. It is very tempting, in Figure 11.11,

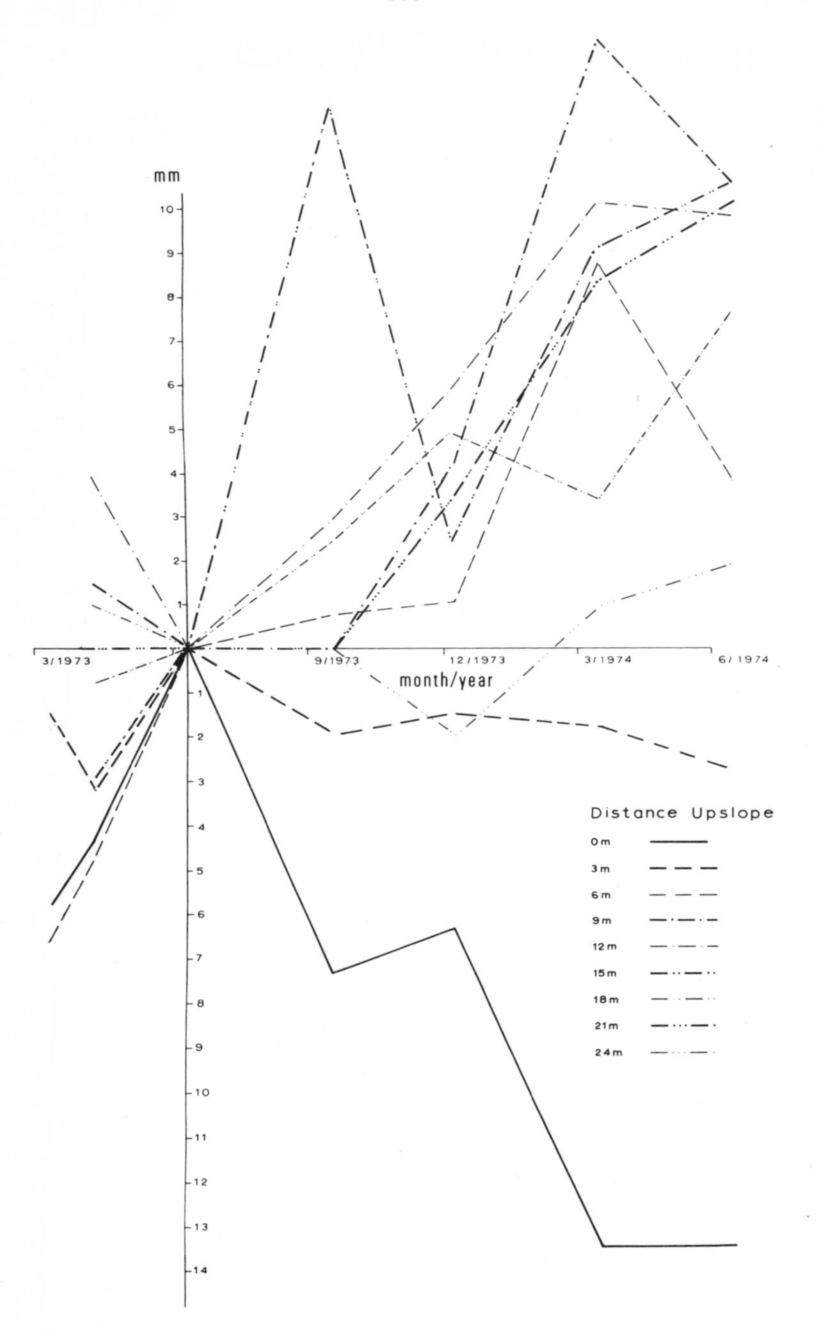

Fig. 11.10.--Milfraen E. Ground Retreat per Grid Row April 1973-June 1974, Independent Slopes. All results set to zero on June 1973. An upward sloping graph records ground retreat. A downward sloping graph records ground advance.

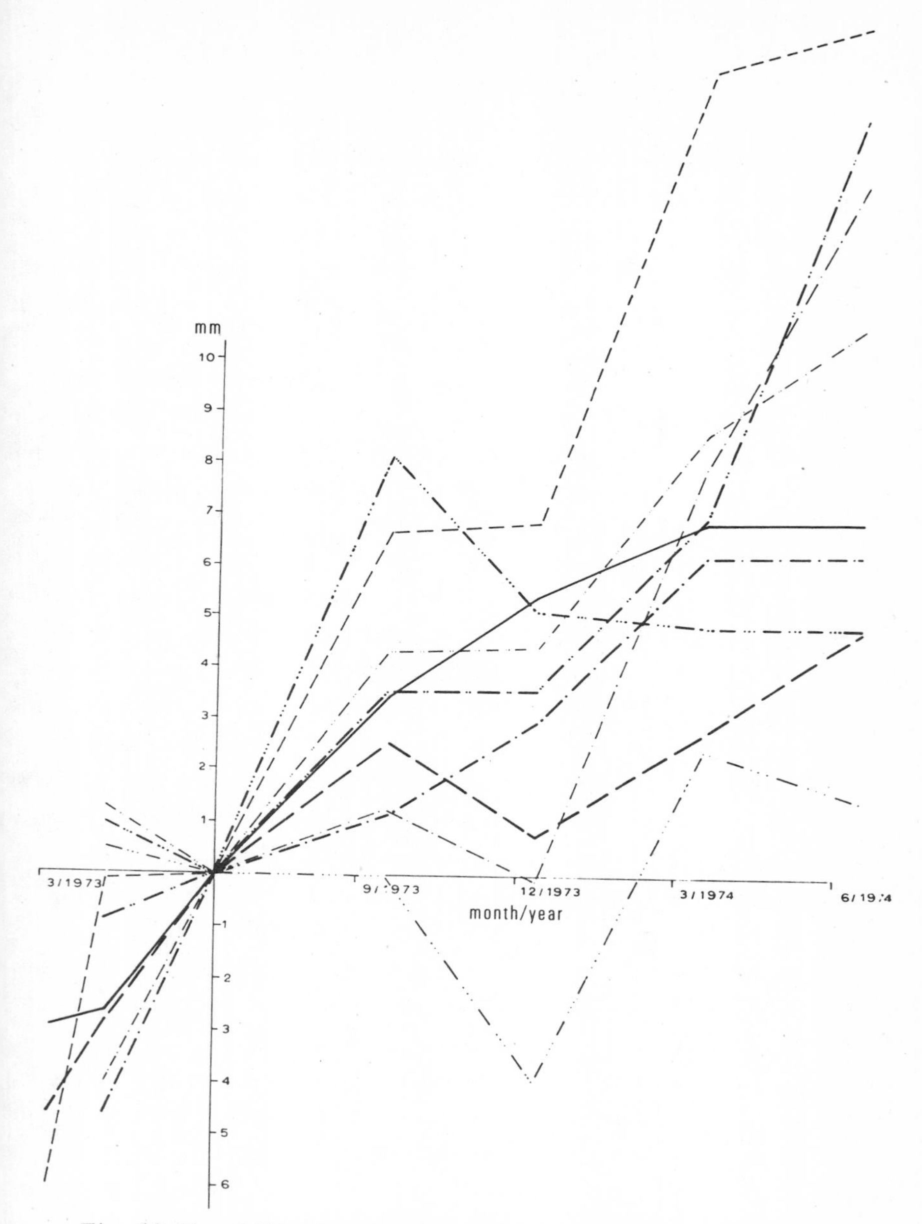

Fig. 11.11.--Milfraen E. Ground Retreat per Grid Row 1972-1974, Gully Controlled Slopes. All results set to zero on June 1973. A rising graph denotes ground retreat.

to attempt to relate the internal variations of ground retreat on the gully controlled slopes to past conditions of erosion or deposition in the gully channel (Figure 11.12). These might be transmitted upslope rather as knickpoints are transmitted upstream. Unfortunately, the data is of insufficient duration to support any test of such an assertion, though the idea just might be tied in with those of Marosi and Szilard (1969; Marosi - 1971) who support the hypothesis of slope development by travelling inflexions. This suggests that slope evolution proceeds by the upslope migration of bands of erosional, depositional, and neutral slope activity. The only reasonable stimulus for the existence of such travelling inflexions is the basal fluvial control of the slope.

The gully channel is the most active profile. However, it is not that which experiences most retreat because erosion is balanced by the deposition of debris from the gully controlled slopes. Activity is most pronounced in the mid slope (12 meters) and declines progressively upslope. This is probably a function of the fact that the lower main slope has the longest run of gully channel which is not influenced by the deposition at the spoil tip's foot, and therefore has the greatest capacity for erosion, while it also has a longer reach of gully-controlled and up-channel slopes from which to derive sediment than do the slopes of the upper main slope and upper convexity. The progressive decline of retreat scores down slope reflects channel aggradation at the junction of the coal spoil slopes with the surrounding moorland. It is possible to assign a numerical value to these variations in activity by the calculation of an "Activity Index." This index is simply the sum of the deviations of a profile's "Q" (quarterly ground retreat estimate) values from zero. The within-profile variation of morphogenetic activity is illustrated by Figure 11.13. However, in spite of this activity the results from the period March 1973-June 1974 testify to only two continuing changes in channel profile. The first is the amelioration of the basal concavity by channel aggradation, and the second is the extension of the upper convexity. It is unlikely to be a co-incidence that the two grid rows which show the greatest net retreat during the period of the experiment are the two which are located on the upper convexity (21 meter and 18 meter). The origins of the out-wash illustrated by Photograph 11.11 are, therefore, not on the most active parts of the channel profile, whose role is mainly transportation, but on the less active upper channel and the gully-controlled slopes.

The "Activity Index" may also be used to assess the relative importance of the gully channel to slope development. The activity estimate for the whole gully channel is 38.3. This means that the gully channel is, per unit area, four

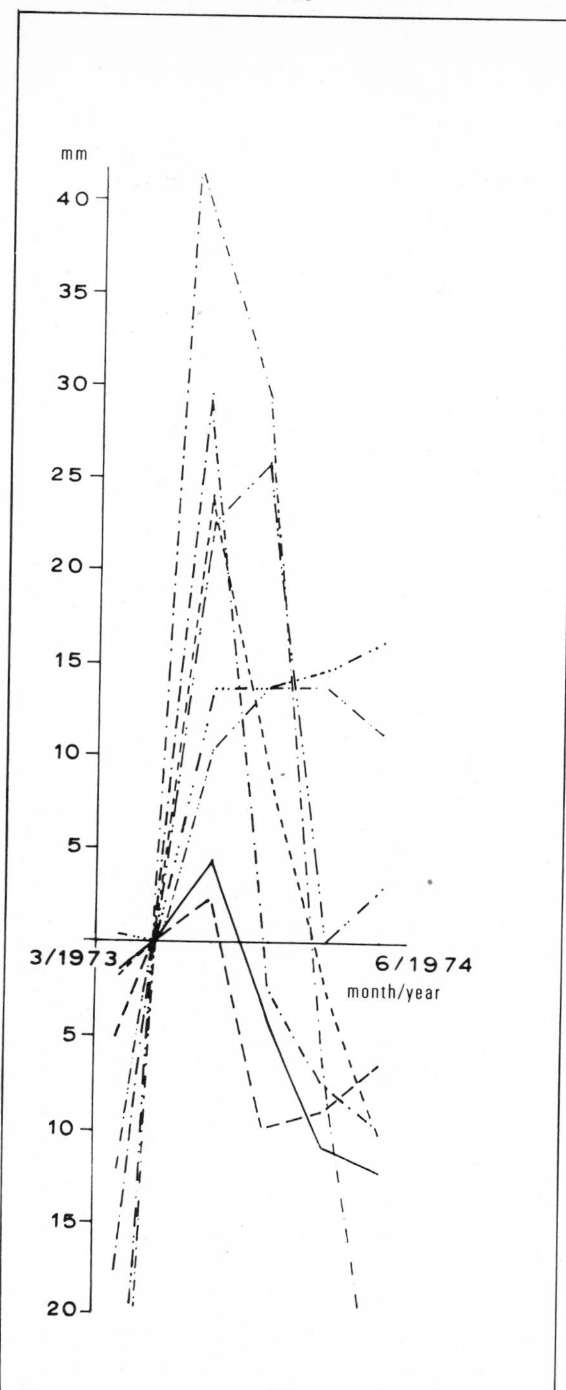

Fig. 11.12.--Milfraen E. Ground Retreat per Grid Row 1972-1974. Gully channel. A rising graph denotes ground retreat. (Legend, see 11.10.)

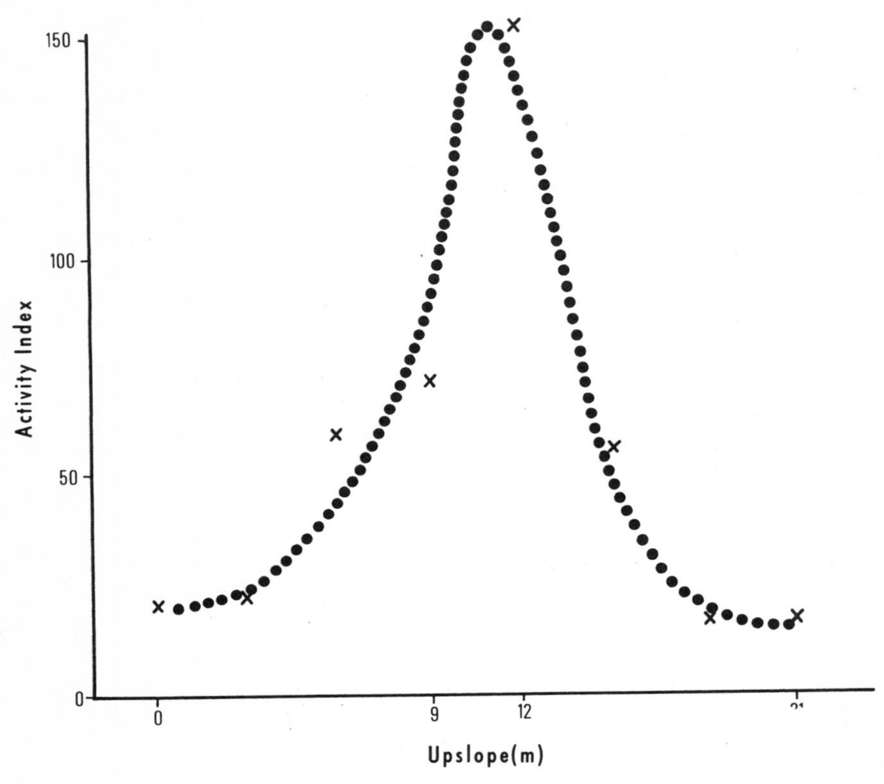

Fig. 11.13.--Variation of Geomorphological Activity within the Gully Channel.

P. 11. 11. --Outwash from Milfraen Spoil Tips

times as active as the gully-controlled slopes (9.4) and over six times as active as the gully-independent slopes (6.1), during the period June 1973-June 1974. A consideration of the activity of the two vegetated profiles, Milfraen S.W. and Milfraen N.E., during the same period, suggests that the south-west facing profile was over six times as active as the north-east facing profile (6.9 vs. 1.1), and indeed that it was more active than the comparable east facing, unvegetated profile of Milfraen E. (6.1). This latter result underlines the frequently underestimated importance of aspect to slope development. The former result emphasizes the pre-eminence of gully channels is effecting denudation, and suggests that the activities of this gully may have done much to accelerate the devegetation of Milfraen E. by increasing erosion and the mobility of the surface layers of the undercut gully-controlled slopes to an extent where the turf cover was broken and destroyed. This devegetation has now spread to affect slopes which are beyond the gully's direct influence. So, if this hypothesis is correct, the morphologically gully-independent slopes of Milfraen E. are not really gully-independent after all, and their development might be explained in terms of devegetation emanating, at least indirectly, from local gully channel development.

It is possible to conceptualize these patterns of ground advance and retreat in terms of slope evolution. Thus, a third series of graphs have been prepared in which ground retreat is plotted along the slope profile. Of course, the ground retreat recorded during the short time-span of these experiments is not of the same order of magnitude as the gross slope morphology and so the actual retreat values have had to be multiplied by a hundred to achieve the visual impact of Figure 11.14 and Figure 11.15. Nevertheless, these graphs serve to underline the main conclusions concerning slope development on the plateau spoil tips at Milfraen.

Figure 11.14 represents separately the patterns of slope evolution suggested by each year's ground retreat records on both of the vegetated slope profiles. It can be seen that morphogenetic changes were concentrated at the slope foot and on the upper convexity. During the two years of direct measurement the dominant mode of slope evolution seems to have been main slope replacement (Carson and Kirkby - 1973) by the extension of the upper convexity, and to a very much smaller extent by the development of slope foot accumulation. These processes are more clearly expressed on Milfraen N.E. The results from Milfraen S.W. contain more than a suggestion of slope decline. Broadly, however, there is a tendency on these two vegetated slopes for current slope

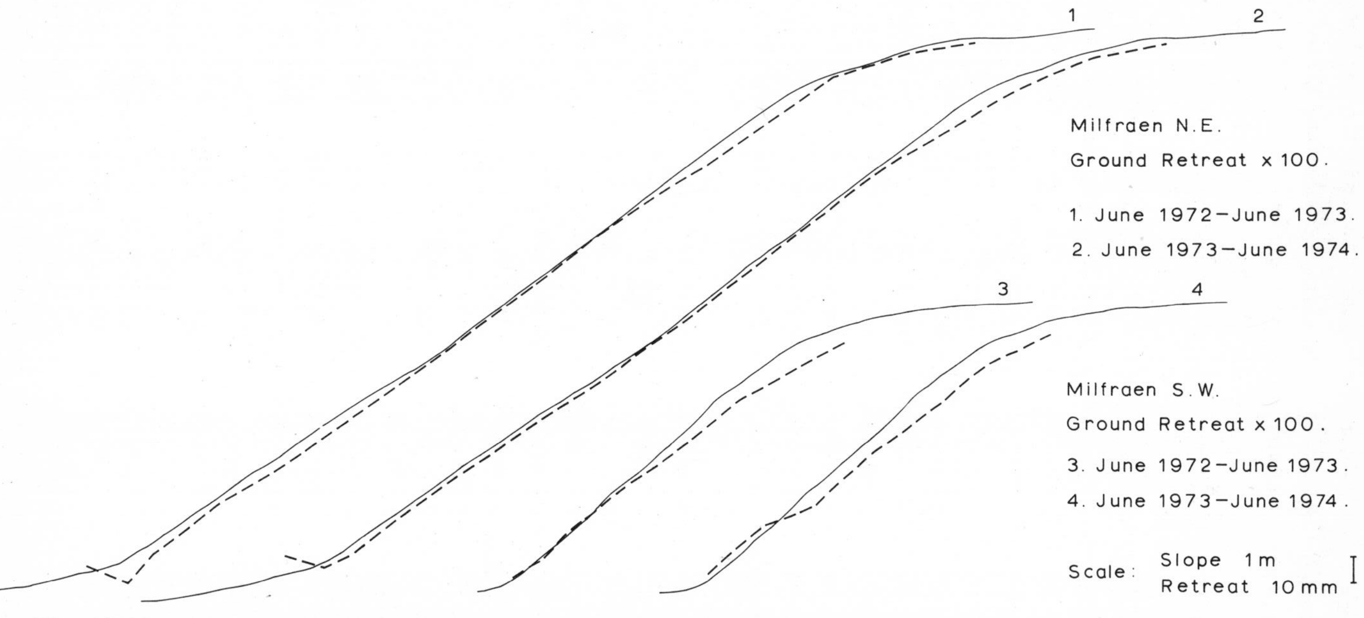

Fig. 11.14.--Milfraen Vegetated Slopes, Ground Retreat 1972-1974. On these graphs ground retreat x 100 is graphed along the actual slope profile. The two years of the experiment are graphed separately.

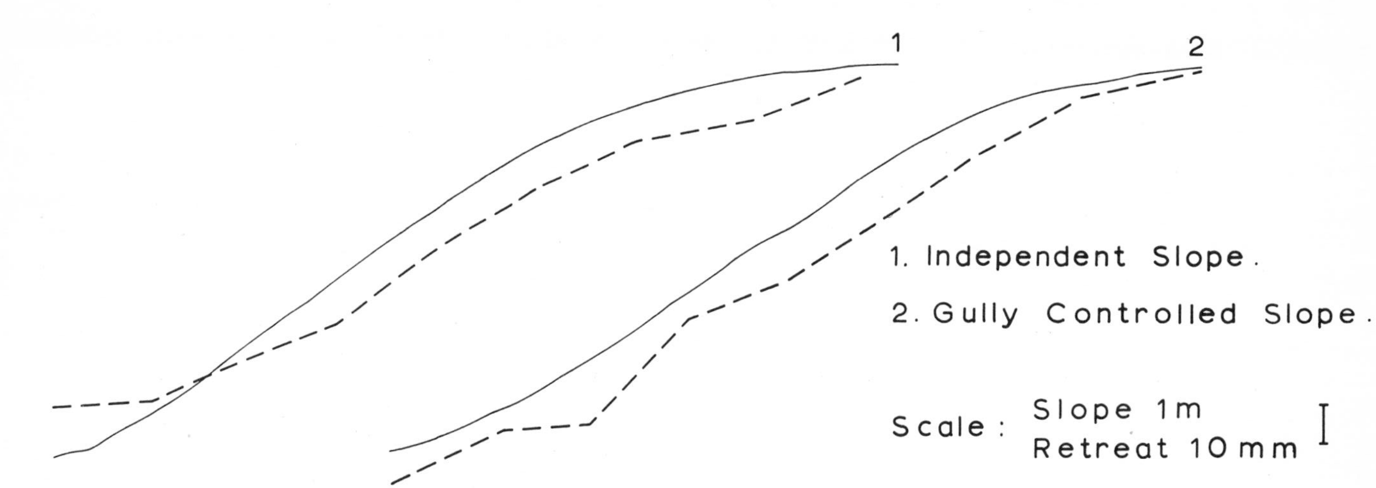

Fig. 11.15.--Milfraen Unvegetated Slopes, Ground Retreat June 1973-June 1974. On these graphs ground retreat x 100 is graphed along the actual slope profile.

alteration to reflect precisely current slope morphology.

Figure 11.15 compares the modes of slope evolution suggested by the ground retreat record from the gully-independent and gully-controlled slopes of Milfraen E. during the single complete year of measurement. The graph from the independent slopes (Figure 11.15.1) is very similar in form to that for Milfraen S.W. in the same year (Figure 11.14.4). However, it clearly reiterates the conclusions offered for the vegetated slopes--that slope evolution proceeds by parallel retreat midslope, the extension of the upper convexity, and the growth of a slope foot accumulation. The graph for the gully-controlled slope again supports the conclusion of parallel retreat midslope but conditioned by the upslope migration of bands of deposition and erosion generated by erosion and deposition in the basal gully channel.

The record from the gully channel itself cannot really be represented in the form of Figure 11.14 and Figure 11.15. However, it may speculatively be asserted that, if this brief record of erosion and deposition is representative of normal conditions, then the main morphological changes accomplished in the gully channel itself are not incision, except on the upper-most part of the thalweg, but headward extension coupled by the growth of an alluvial fan at the slope foot.

To summarize: this study has considered the effects of three important controls on ground retreat and slope development: aspect, vegetation and gully development. It was discovered that the south-west facing slope of Milfraen suffered a greater ground retreat in both years of the experiment than did the slope which faced north-east: 5.8 mms vs. 2.8 mms in the period June 1972-June 1973, and 3.5 mms vs. 0.4 mms in the period June 1973-June 1974. The average mean retreat for the vegetated slopes in two years of direct measurement was 3.1 mms p.a. Mean ground retreat recorded for the unvegetated slopes of Milfraen E. during the period June 1973-June 1974 was: 4.6 mms on the interfluve (gully-independent) slopes, and 8.5 mms on the slopes which had a basal gully control. Slope evolution was dominated by main slope replacement by the extension of the upper convexity.

CHAPTER 12

GROUND ADVANCE ON THE LOWER SLOPES OF THE LOOSE-TIPPED SPOIL CONES AT BIG PIT AND KAY'S SLOPE

The two tips selected for these experiments are both typical MacClane Shute coal spoil cones. They were created by the loose-tipping of coarse, irregularly sorted, shaley discard from an extending conveyer belt laid on one side of the coal spoil tip. The tips have a local relief of around 40 meters, their maximum slope angles are between 34° and 36° and these are controlled purely by the angle of internal friction of the dumped spoil. The tips' surfaces are very unstable and prone to sliding. They support very little vegetation and are dissected by gullies which achieve a vertical depth of almost 1 meter at Big Pit Tip. Both tips are products of the twentieth century. The spoil cone at Big Pit was created between 1935 and 1964 (Photograph P.12.1). Tipping at Kay's Slope (Photograph P.12.2) commenced in 1943 and ceased when the mine was abandoned in 1965. Neither tip has suffered combustion.

Erosion grids were established on each of these tips during June and July 1972. Because of the coarse, scree-like nature of much of these spoil tips' material, and because of the great instability of the steep coal tip slopes, the sites for the location of the erosion pins had to be carefully chosen. Erosion pins are inappropriate to studies of the movement of very coarse materials, so instrumentation had to be limited to those parts of the tip where fine spoil predominated. The instability of the spoil tips' surfaces meant that greater than usual care had to be taken during establishment and data collection to avoid causing a disturbance of the spoil in the immediate vicinity of the erosion pins. In practice, disturbance could only be limited by limiting instrumentation to the lower and less steep slopes of the spoil cone.

The data collected from these tips are blemished by human interferences. Three erosion pin grids were established but only one survived the course of the experiment. One grid, established on the south-western slopes of the Kay's Slope cone, was completely destroyed by vandals twice during the first six

P. 12. 1. --Big Pit Tip View up MacClane Shute

P. 12. 2. --Kay's Slope MacClane Shute Cone--Erosion grid sited to left of telegraph pole

months after establishment and this site had to be completely abandoned. A second grid, established on a south facing slope beneath the head of the MacClane Shute on Big Pit's Tip (Figure 12.1), operated undisturbed, but only until April 1973 when this tip was levelled by land reclamation work. The third grid, which was established on the north-east facing slopes of Kay's Slope's Tip (Figure 12.2), lost a third of its total stock of erosion pins to vandals, but, alone, survived the two years of the experiment. The grid was finally destroyed during the construction of an access road on the commencement of the working of the tip by Ryans Holdings Ltd. (Cardiff), in the summer of 1974 (P.12.2).

The results of the experiments are summarized in Figure 12.3. The results from Big Pit's Tip (Figure 12.3a) are of too brief a duration to do much more than lend support to the observations on Kay's Slope's Tip (Figure 12.3b). However, it can be seen that during the period June 1972-June 1973, there was a prolonged and active deposition of debris on most of the lower slopes of both tips. This amounted to an average ground advance of just over 10 mms on the instrumented sections of the slope. It is interesting that on both graphs one trace suggests a condition of comparative neutrality. In Figure 12.3a, the 6-meter trace, and in Figure 12.3b, the 3-meter trace, are both graphs which are markedly different from those from other sections of the lower slopes. It is possible that these records testify to the development of a transportation slope behind an advancing front of slope foot deposition. This suggestion is supported by Figure 12.4. in which the average annual ground advance, multiplied by a factor of fifty, is graphed along a mean flow line slope profile of the lower slopes of Kay's Slope's Tip.

It is unfortunate that the ground advance records from each of the two years' observation of Kay's Slope Tip are so different. The ground advance recorded in the second year of observation is less than a third of that recorded in the first year. Further, the pattern of advance is completely different. The 0-meter grid row trace which recorded a massive 20.0 mms advance in the period June 1972-June 1973, recorded a slight retreat in the following year. Similarly, the 9-meter trace which recorded a 15.4 mm advance in 1972-1973, recorded only a 2.5 mm advance in the following year. Thus, it is the 15-meter and 3-meter traces which recorded the smallest advances (5.4 mm and 1.0 mm only) in the first year of observation, but which record the largest advances in the second year: 7.1 mm and 4.3 mm respectively. It is probable that many of these differences can be attributed to differences in the morphogenetic regimes of each year. However, it is possible that some of these varia-

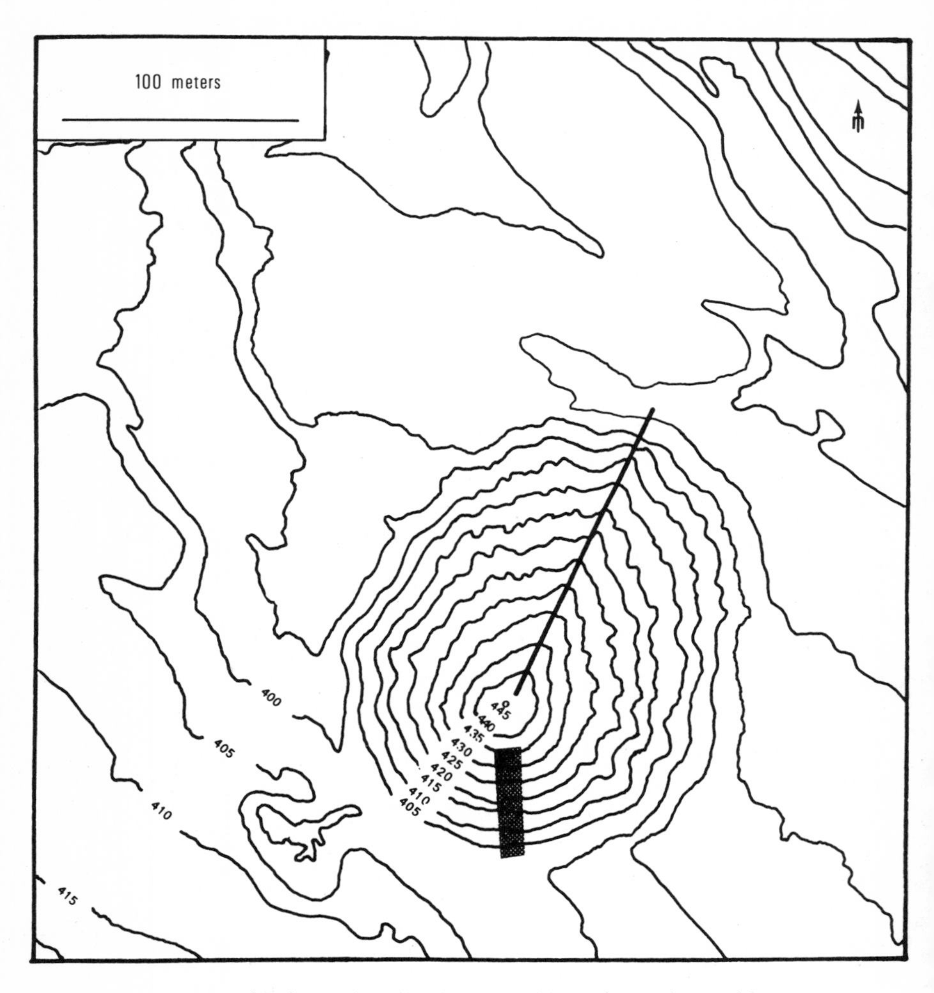

Fig. 12.1.--Big Pit Tip--Location of erosion grid

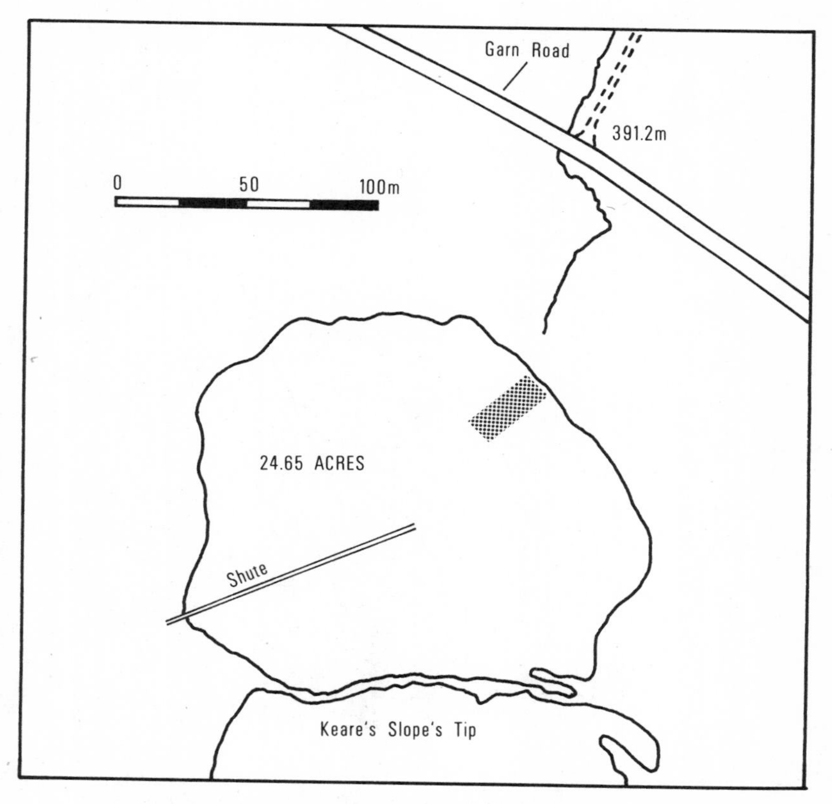

Fig. 12. 2. --Kay's Slope Tip--Location of erosion grid

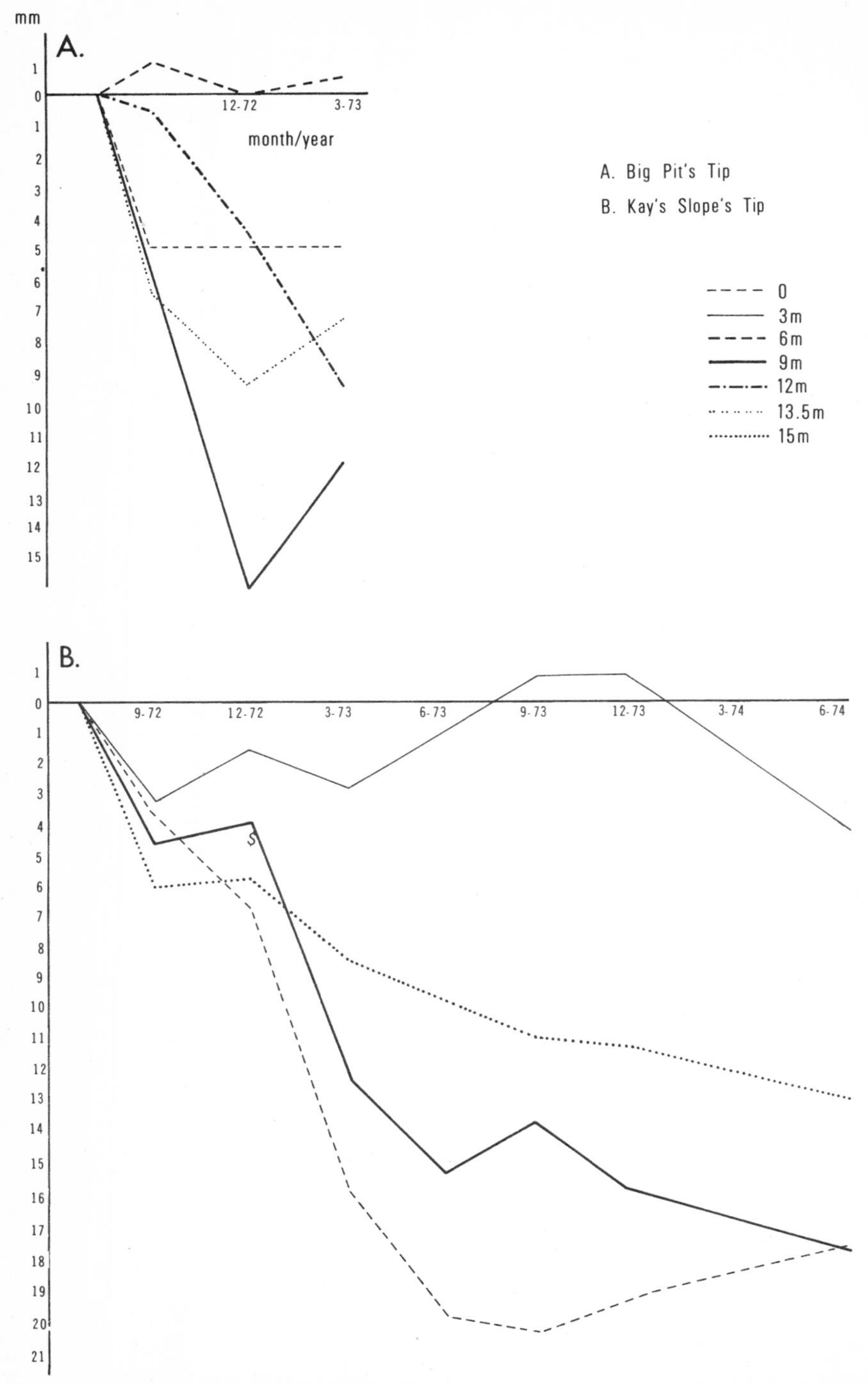

Fig. 12.3.--Ground Retreat per Grid Row, Lower Slopes of: A. Big Pit's Tip; B. Kay's Slope's Tip.

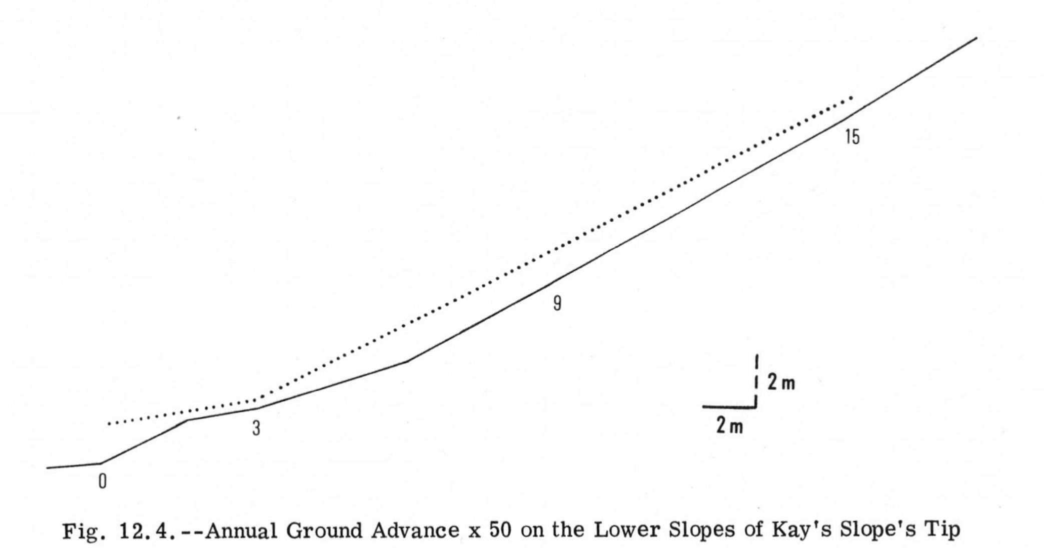

Fig. 12.4.--Annual Ground Advance x 50 on the Lower Slopes of Kay's Slope's Tip

tions are due to the activities of soil creep which had tipped the erosion pins on the upper two grid rows of Kay's Slope's Tip through 45° by June 1974. In any case, this inconsistency in the experiment's results must constitute a major cause for concern.

The experiments established on the MacClane Shute cones at Blaenavon are thus among the least satisfactory areas of this study. The erosion pin technique is inappropriate for studies of steep unstable surfaces and coarse materials. Because of this these experiments have had to be restricted to the lowest slopes and to areas with atypically fine spoil. All of the experiments were hampered by the loss of equipment through vandalism, land reclamation works, the movement of large blocks of debris and small scale slides. Less than 20% of the erosion pins installed during the summer of 1972 survived into the summer of 1974. Further, of those which survived, 50% had been severely tilted by soil creep which probably makes a significant impact on the second year's data records.

However, it might be suggested that these measurements do have some redeeming features. First, they are the only records which have ever been made of the morphometric evolution of these locally prominent land-forms. Second, the rapidity with which these landforms are disappearing to land reclamation works makes it probable that these records are the only measurements which will ever be made of the rates of evolution of MacClane Shute Spoil Cones. Third, the results themselves are not entirely uninteresting. They suggest that the lower slopes of these spoil cones are aggrading at a rate of between 5 and 8 mms p.a. and give evidence of the development of a transportation foot slope separating the main lower concavity of the cone from a front of advancing spoil accumulation.

Mizutani (1969, 1970) has drawn comparisons between the evolution of colliery spoil cones and certain volcanic cinder cones for the purposes of some mainly theoretical discussions of slope evolution. It will be interesting to see if future studies of micro-erosion processes and their effect on the evolution of volcanic cinder cones yield results which are comparable to those gathered on the artificial landforms of Blaenavon.

CHAPTER 13

GROUND RETREAT AND SLOPE DEVELOPMENT ON COLLIERY SPOIL MOUNDS

Colliery spoil mounds are among the most obtrusive of artificial landforms. However, at present, relatively little is known about the geomorphic evolution of these features after their construction, or of their contribution to the problems of sediment pollution. However, it would seem that the coal mining industry is with us to stay and it is likely that coal tips and the landforms subsequently derived from coal spoil in the process of reclamation will become increasingly common in the years to come.

This study has been concerned with the evolution of slope and gully profiles on two fundamental classes of colliery spoil mound--the loose tipped spoil cone and the smaller features created by tipping and compaction. Direct measurements of ground retreat and slope evolution have been gathered over a period of two years from instrumented slope profiles established on the lower slopes of two large spoil cones and three lobes of a compacted tip of the "plateau" type (Jones et al. - 1972). The evolution of slopes on the other common types of compacted spoil tip--the nineteenth century "fan-ridge" spoil mounds and the twentieth century "land reclamation project" slopes--have been examined by morphometric analysis.

The results from the less successful examination of ground advance on the lower slopes of the loose-tipped spoil cones suggest that the rate of accumulation of these features is proceeding at a rate of between 5 mms and 8 mms p.a. It was revealed that an important feature of the morphological development at the slope foot was the development of a transportation foot slope separating the main lower slope element from an advancing front of spoil accumulation. Slope evolution is influenced by the surface instability of these features. The importance of creep and small scale failures and sludgings of the surface debris makes these tips a less than suitable subject for examination by the erosion pin technique.

Instrumented slope profiles were also established on three lobes of a compacted plateau-type spoil tip complex. The instrumented slopes on the loose-tipped spoil cones were unvegetated and unaffected by gullying. Here, however, it was possible also to study slopes which were vegetated and of opposed aspect, and slopes which possessed a basal gully control. The results from the unvegetated interfluve slopes suggest that the average annual rate of retreat is 4.6 mms p.a. The results from the two vegetated profiles suggest that these slopes retreat at a much smaller rate--3.1 mms p.a. In both years of measurement, it was found that the vegetated profile which had a south-west aspect (Photograph P.13.1) retreated faster than that which faced north-east. The greatest rate of retreat recorded at Milfraen was discovered on the sections of the unvegetated slopes which were affected by a basal gully control--8.5 mms p.a. The most active profile examined was that of this basal gully channel, but here, erosion was largely balanced by deposition, presumably, the deposition of the debris removed from the gully controlled slopes.

The morphometric examination of older, compacted, fan-ridge, tips confirmed that these features may be regarded as a single data population. The average maximum slope angle for these tips was about 33° and the standard deviation about this mean was around 4°. Maximum slope angle was unaffected by slope height or aspect. Measurements were also made of slope profiles on a modern compacted tip and a modern reclamation. It was found that the major morphometric difference between the slopes of the modern and nineteenth century landforms was that the modern slopes tended to have an abrupt junction between the main slope and crest units while the older tips possessed a very well developed upper slope element. It was concluded, therefore, that the dominant mode of slope evolution on artificial landforms constructed from colliery spoil materials was the replacement of the main slope segment by the development of a well developed upper slope element. This observation was supported by particle size analyses of surface spoil at different points on the slope profile. The spoil at the surface on the more active upper element of the compacted tips was far more coarse than that from other parts of the slope profile. Further confirmation comes from an examination of the results from the instrumented profiles on the plateau spoil at Milfraen which tend also to show that it is the upper slope element which suffers the greatest ground retreat.

Devegetation is a persistent problem for land reclamation projects and seems to be a widespread feature of these compacted colliery spoil mounds. The initiation of devegetation on these tips seems frequently to be associated with the activity at the upper convexity. There seems to be a tendency for rup-

P. 13. 1. --Southern Slopes of the Milfraen Spoil Tip Complex. View south to the loose-tipped MacClane Shute spoil cones at Kay's Slope and Keare's Slope across Waunafon.

P. 13. 2. --Sheep Trampling and Burrowing near Milfraen E. Erosion Grid

tures in the vegetation to develop at these locations. The importance of these ruptures may be exacerbated by the trampling and "burrowing" activities of sheep (Photograph P. 13. 2). In any case, it seems that these bare scars of the upper convexity are capable of extending themselves, upslope by undermining the turf of the slope crest, downslope by burying the thinner mid-slope turf in loose fragments. A survey of compacted spoil tips in the Garn Road area of Blaenavon demonstrated that while 42% possessed a complete vegetation cover, 34% had developed bare scars at the upper convexity and on a further 17% this scar had extended to affect the whole slope. Slopes which had a downslope divergence of flow lines seemed especially susceptible to devegetation.

CHAPTER 14

THE INFILLED OPENCAST (SURFACE MINE) SITE ENVIRONMENT (REVIEWED)

The essential differences between the erosional environment of an infilled opencast pit and a deep mine spoil tip are composition and scale. The composition of a colliery spoil tip is controlled mainly by the nature of the coal seam and the quality of the related strata. The opencast fill, however, consists of a jumbled mixture of all the strata overlying the coal seam. It thus contains a far higher proportion of non-carboniferous debris and a broader spectrum of non-coal-shale strata. Again, deep mine spoil tips tend to be small discrete landforms which occupy not more than a couple of dozen acres. An opencast mine is a much larger operation and opencast mine spoil banks may extend over hundreds of acres. This is important because considerations of deep mine spoil banks must inevitably be concerned with the external faces of these features and their relationships with the surrounding landscape. Discussions of opencast-mine spoil banks, however, are more likely to be concerned with the internal relationships of the entirely artificial overburden landscape.

Opencast operations are a relatively small part of the concern of Britain's National Coal Board. In 1963 production from deep mines in the South Wales Coalfield was 18.89 million tonnes (18.6 million tons) while opencast mining produced only 1.42 million tonnes (1.4 million tons) (Morgan - 1965). This production ratio is not reflected in other parts of the world. In the United States, 39% of the national output is from surface coal workings (Rushton - 1973) while in the U.S.S.R. opencast mines produce about 28% of all coal extracted (Brealey and Atkinson - 1968). In the free world as a whole during 1965, 26% of all non-metalliferous minerals, and 61% of all metalliferous minerals, were produced from open-pit operations (Anon. - 1967), and the scale of these operations increases each year. In the U.S.S.R., the average annual per-coal-strip-mine production was 1.75 million tonnes in 1956 but about 3.2 million tonnes during 1966-1970. In the United States, 3.2 million acres have been disturbed by surface mining, 35% by open-pit operations, and 56% by strip-mining (U.S. Department of the Interior - 1967).

All available evidence suggests that opencast mining is an expanding activity. Brealey (1966) has compiled a list of 13 factors which favor opencast against underground mining:

(1) Higher productivity;
(2) Greater concentration of all operations and simplified management of men and machines;
(3) Greater output per mine;
(4) Lower capital cost per ton mined per annum;
(5) Lower operating cost per ton mined;
(6) Possibility of moving a higher ratio of waste to mineral and the exploitation of lower grade reserves;
(7) Greater geological certainty and easier exploration;
(8) Less limitation on size and weight of machines;
(9) Greater simplicity in auxiliary operations and services;
(10) Greater recovery of mineral and less dilution;
(11) Greater reserves;
(12) Simplified engineering, planning and control;
(13) Greater safety.

The techniques of surface mining are not strictly relevant to this account except as they affect the mixing and surface structures of the spoil. The different methods of surface mining are described and illustrated by the U.S. Department of the Interior Report, Surface Mining and Our Environment (U.S. Dept. Interior - 1967). The more common schemes of operation and their economic limitations are discussed by Brealey and Atkinson (1968) and Rushton (1973) and these techniques are classified and illustrated as Figure 14.1 and Figure 14.2.

In practical terms, however, the principles and processes are much the same for all types of opencast and strip mine operation. First, top-soil and sub-soil to a depth of up to a meter--but usually no more than 200 mm--is removed and stacked on the pit's fringes. Then, the overburden is removed in 20-30 meter strips and temporarily dumped while the coal seam beneath is excavated. The coal itself is then removed to a screening plant where it is processed. The screenings are usually dumped as a small fan-ridge tip which may be adjacent to the opencast workings. After the coal has been excavated the overburden is pushed back into the pit or trench. Eventually, usually after the mine is closed, the dumps are regraded, the by now rather mixed top-soil and sub-soil horizons are replaced, and in the face of the rampant surface erosion of all but the most gentle slopes, an attempt is made to re-establish vegetation

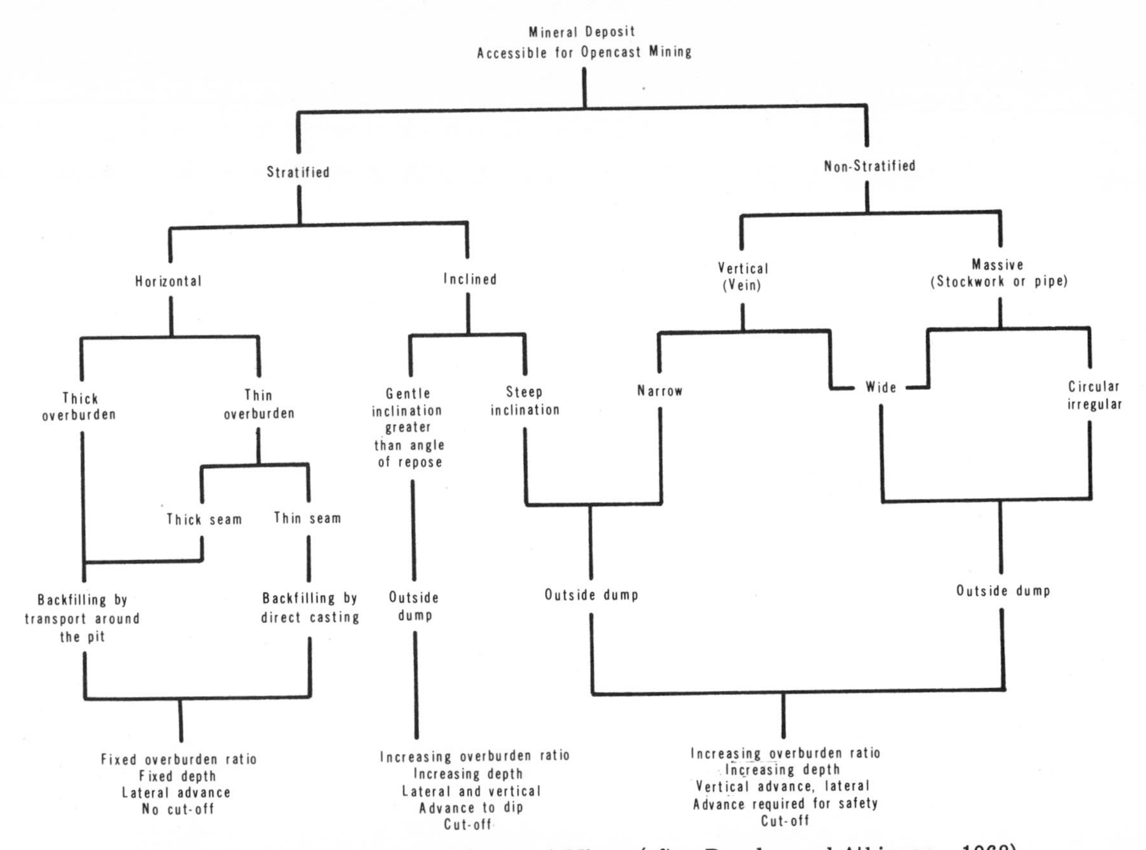

Fig. 14.1.--Classification of Opencast Mines (after Brealey and Atkinson - 1968)

on the disturbed land-surface. Normally, the land is then left to settle for several years. Chwastek (1970) has suggested that finally land drains are installed and the land is returned to profitable agricultural uses. In most modern British opencast operations, all these processes are rigidly adhered to and, frequently, the result is agricultural land of improved quality (Davison - 1974). However, some older sites, especially those worked in the 1943-1950 period were never properly restored (Thomas - 1966), and there are some later sites, especially those located on exposed slopes in mountainous areas, where reclamation has failed to promote sufficient vegetation growth to prevent disastrous erosion.

During a four-year study of the sediment yield from Kentucky spoil banks, the average annual sediment yield was calculated as 27,000 tons per square mile. This was over one thousand times greater than that from comparable forested areas in the same period (25 tons per square mile p.a.) (U.S. Dept. Interior - 1967).

A survey of erosion problems on U.S. surface mine spoil banks discovered that while there was no serious erosion problem on 60% of the sites examined. Most of the remainder possessed small gully channels and 10% possessed gully channels which were more than 300 mms deep. Sediment pollution was a feature of 56% of the ponds and 52% of the streams adjacent to the surface mining spoil banks (U.S. Dept. Interior - 1967).

Naturally, the mechanical and chemical properties of overburden are highly variable (Fitton et al. - 1959). However, most overburdens tend to be very acid. In the United States, 47% of surface mined spoil banks have a pH of between pH 3.0 and pH 5.0, and 1% have an average pH of less than 3.0. This acidity causes the chemical pollution of local water bodies--mainly through the addition of sulphur compounds--and inhibits the plant colonization of the actual spoil bank sites. In the United States, only 15% of all surface mined spoil banks support an adequate vegetation cover, though a further 15% support a cover which is sufficient to prevent the worst excesses of erosion. The absence of vegetation on the remaining 70% of surface mined spoil banks represents a massive problem of landscape disfiguration by the evolution of badland type topography and the spread of, sometimes toxic, sediment outwashes (U.S. Dept. Interior - 1967).

The conditions of badland-type landscape evolution on a typical surface mine infill were described by Schumm (1956a).

The Perth Amboy claypit is located in New Jersey on the banks of the

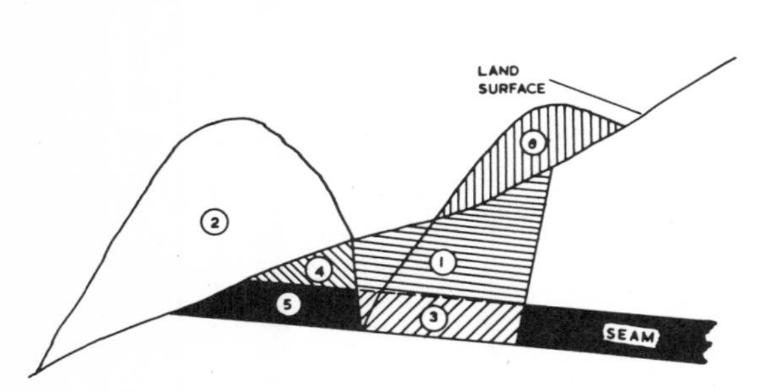

SEQUENCE OF OPERATIONS

1. A BOX CUT SECTION OF OVERBURDEN, AREA No. 1, IS REMOVED AND DEPOSITED IMMEDIATELY DOWN THE SLOPE AT AREA No. 2
2. THE EXPOSED COAL AT AREA 3 IS THEN EXCAVATED AND TRANSPORTED AWAY
3. THE OVERBURDEN AT AREAS 2 AND 4 IS THEN TRANSFERRED TO AREAS 1, 3 AND 6
4. THE EXPOSED SEAM AT AREA 5 IS THEN EXCAVATED AND TRANSPORTED AWAY
5 THE OVERBURDEN AT AREA 6 IS THEN LEVELLED OFF TO PROVIDE AN EVEN SURFACE WHICH MAY BE RESTORED TO THE ORIGINAL CONDITION

Line Diagram to show Sequence of Operations used in Surface Contour Mining

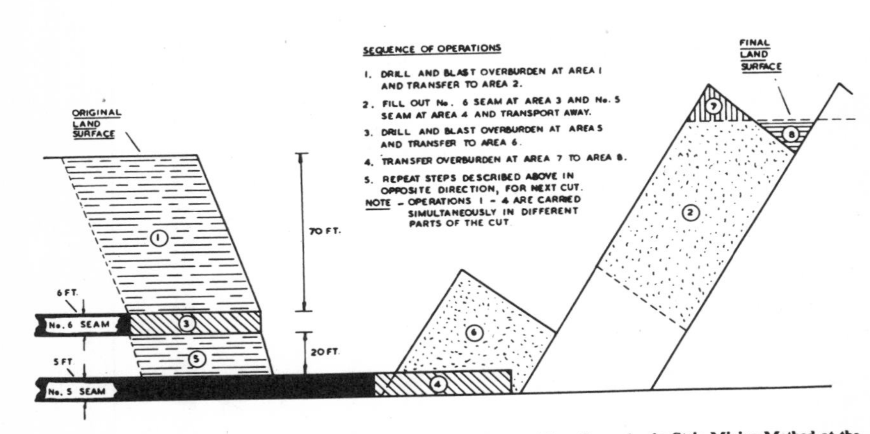

Line Diagram to show the Method of Working for the Extraction of Two Seams by the Strip Mining Method at the South Western Illinois Coal Corporation's Captain Mine

Fig. 14.2.--Methods of Coal Extraction by Surface Mine Operations in the United Kingdom and the United States (after: Rushton - 1973; Brealey and Atkinson - 1968).

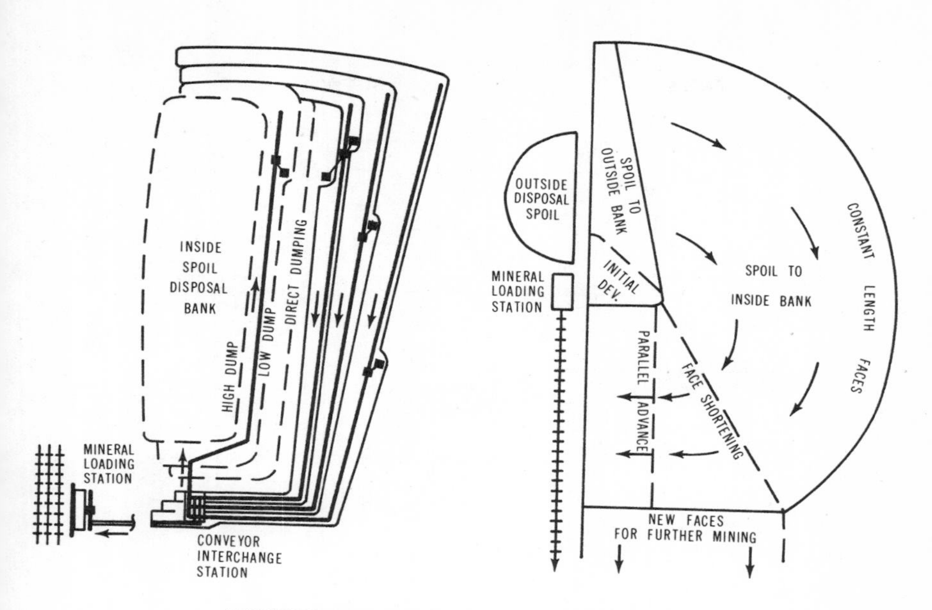

HORIZONTAL DEPOSIT - BELT CONVEYOR TRANSPORT

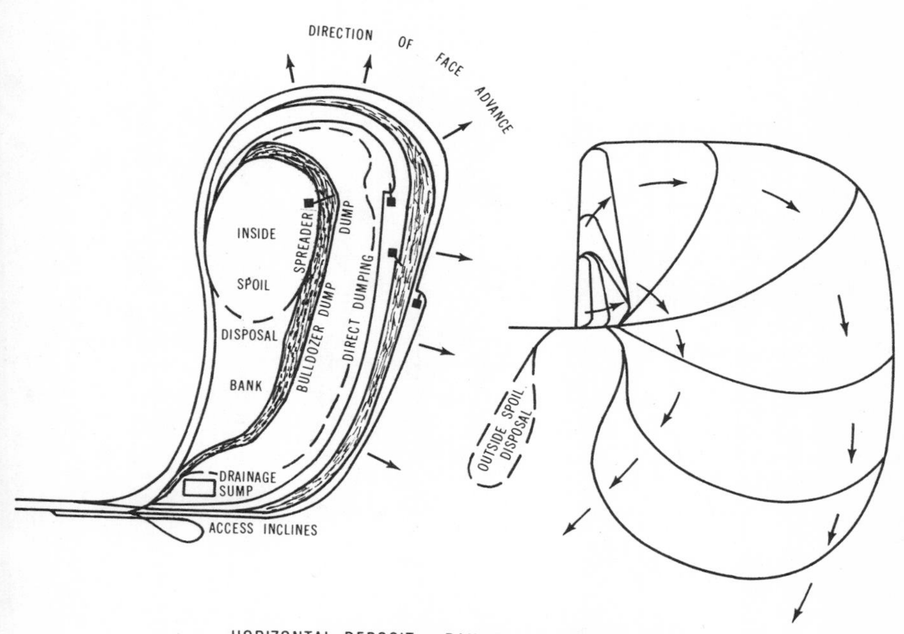

HORIZONTAL DEPOSIT - RAIL TRANSPORT SYSTEM

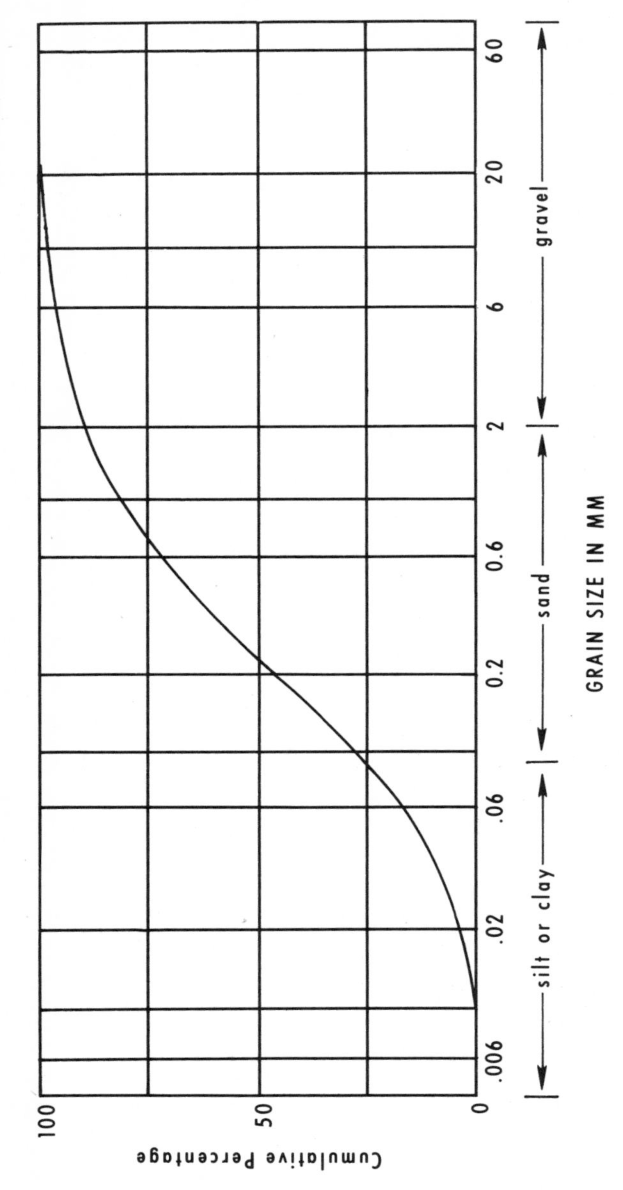

Fig. 14.3.--Grain-size Distribution of Perth Amboy Fill

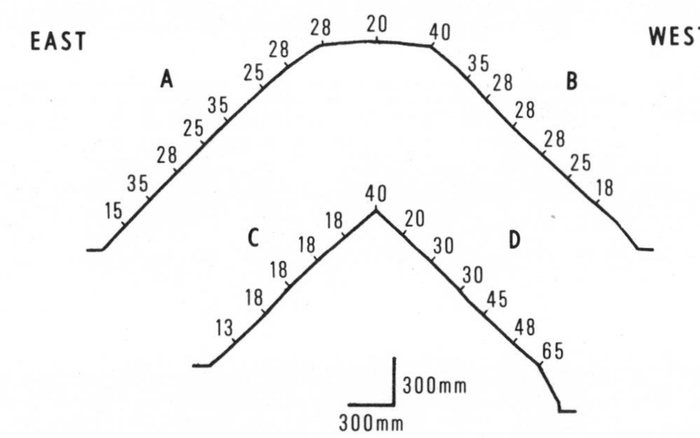

Fig. 14.4.--Erosion of Perth Amboy Claypit Spoil Banks. Tics on profiles show position of stakes; numbers indicate depth of erosion in millimeters. Profiles A, B, and C are typical of the area; profile D shows a basal convexity. (Schumm - 1956a).

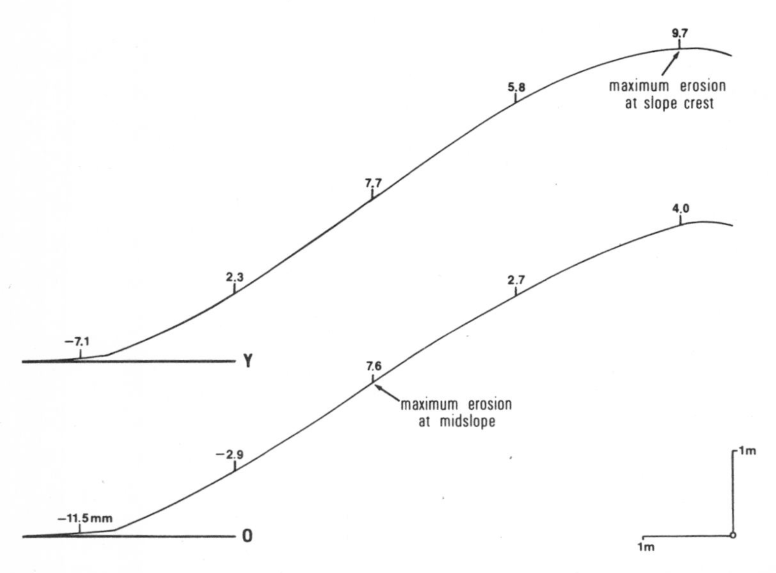

Fig. 14.5.--Erosion of Strip Mine Spoil Banks near Henryetta, Oklahoma. Tics on profiles show position of stakes, numbers indicate depth of erosion in millimeters (Haigh - 1977).

Raritan River. In 1929, this abandoned pit was back-filled with waste and overburden from other pits to create a 12-meter terrace. The Perth Amboy spoil is essentially homogeneous. Its effective engineering size (10%) is 0.04 mms (Smith - 1971). The particle size distribution of this spoil is displayed as Figure 14.3.

The Perth Amboy climate is fairly severe. Temperatures range from -18°C to +35°C. The area suffers a mean annual precipitation of 1072 mms p.a. The site is almost completely devoid of vegetation.

Ground retreat was recorded by means of wooden stakes established as transects on flow line slope profiles. The stakes employed were 457 mm long and 6.4 mms thick, and established flush to the ground surface at 305 mm intervals on the profile. Measurements were recorded to the nearest 2.5 mms. Sixteen profiles were established between June 15 and July 1, 1952. The last measurement was recorded on September 10, 1952. The period is equivalent to a quarterly unit period in the Blaenavon study and most comparable with the summer season records.

Results from the 10-weeks observation demonstrate that erosion was highly variable from point to point along the slope. The greatest ground retreat was recorded on sharp-crested divides and least on convex divides. The retreat of slope segments was essentially uniform and equivalent to circa 23 mms. Figure 14.4 displays results from four of Schumm's profiles (Schumm - 1956a). Figure 14.5 displays some preliminary results from some of the author's studies on surface mine spoil dumps in Eastern Oklahoma. The results, collected on two profiles, by means of rows of 600 mm steel erosion pins over a period of 18 weeks, essentially echo those gathered in Schumm's pioneer study though the totals of erosion are much smaller, a fact possibly relatable to differences in the ages of the sites (Haigh - 1977).

CHAPTER 15

GROUND RETREAT AND SLOPE DEVELOPMENT AT CEFN GARN-YR-ERW ON THE INFILLED OPENCAST SITE AT WAUNAFON

The open-pit mine at Waunafon was the largest of the surface mining oper- ations near war-time Blaenavon. The mine was opened in 1942 and, with the aid of imported American machinery, continued to operate until 1947. Shortly afterwards, the mine was closed and the pit infilled. There is some evidence that the coaling contractor made some attempt to resculpture the surface of the opencast fill and to restore some of the former water courses. However, this restoration is not recognized by today's N. C. B. Opencast Executive which has inherited responsibility for the site. The reclamation could not, in any case, be counted successful since a large part of the site is completely devoid of vege- tation, finely dissected by gullies, and in fact has reverted to an appearance more commonly associated with the "badlands" of North America. The whole area is presently scheduled for reworking and reclamation.

The Waunafon site lies between 380 and 430 meters at the head of the Afon Lwyd valley and between the towns of Blaenavon and Brynmawr. The site occu- pies an area of about 1 km^2, and has an internal local relief of <u>circa</u> 15 meters. The configuration of the spoil is illustrated by Figure 15. 1. The Waunafon Opencast occupies a very exposed site. Its slopes constitute the present inter- fluve separating the shallow Afon Lwyd valley from the Clydach Gorge and the edge of the South Wales Coal-field scarp. There is no record of the undoubt- edly very high wind velocities recorded here, but the Llanelly Hill rain-gauge, located at the site's perimeter, records a heavy and persistent rainfall which averages 1430 mms p. a. The monthly totals of the rainfall recorded during the period of this experiment are graphed as Figure 15. 2.

Approximately half of the site is well to moderately well vegetated. The main areas of vegetation lie to the east and north of the artificial valley of Cefn Garn-yr-erw. On these parts of the site it is noticeable that the plant cover is best developed on the slopes which are least steep. The plant cover is dominated by the grass species *Nardus stricta* L., *Festuca ovina* agg., *F. rubra* L., and *Agrostis tenuis* Sibth. Among the ruderals, thistles are common, especially *Cirsium* spp., and *Sonchus* spp., and so are *Tussilago farfara* L. and *Taraxacum officinale* Weber. Other prominent angiosperms include *Juncus* sp., *Cerastium* sp. and *Empetrum nigrum* L. Mosses and lichens are also well represented.

The spoil at Waunafon is not well mixed but it has a certain homogeneity. The average particle size distribution of representative samples of spoil drawn from several sites at a mean depth of 0.25 meters is displayed as Figure 15.3. It may be assumed that there is quite a wide envelope of variation about this graphed distribution. Nevertheless, it is suggested that in engineering terms (Smith - 1971) the spoil at Waunafon might be regarded as behaving like a fine to medium sand (D^{10}: 0.2-0.6 mms).

Ground retreat is being monitored on a total of four slopes and one gully profile at Waunafon. All of the instrumented slopes lie within the confines of the reconstructed valley at Cefn Garn-yr-erw (Photograph P.15.1) which is the main drainage system of the site. The instrumented profiles have been sited in an attempt to examine the importance of vegetation, variations in the activity of basal gully control, and the development of secondary slope gullies to rates of erosion and the nature of slope evolution. Two of the instrumented slopes are located on vegetated slopes of "opposite aspect." One, Waunafon E. faces east, and the other, Waunafon W., which is located on the other slope of the valley, faces west. A further two instrumented slope profiles, Waunafon D and Waunafon C, are established on a devegetated, west facing slope, which is being undercut vigorously, but to varying degrees, by the channel in Cefn Garn-yr-erw. One of these profiles, Waunafon C, is established adjacent to a gully channel which is also being monitored. The instrumented gully profile is referred to as Waunafon B. The locations of all these experiments are displayed in Figure 15.4.

Two types of instrumentation are employed by the experiments at Waunafon. All the experiments are based on the use of two lengths of erosion pin: 610 mm and 153 mm. These pins are established in fixed format rows down the flow line slope of the instrumented profile. Fixed format rows are one meter wide and bounded by two 610 mm erosion pins. Three 153 mm pins are spaced out at 250 mm intervals between. The shafts and tops of the pins are aligned so that movements and disturbances affecting individual pins may be readily apparent. The pins are allowed an initial exposure of *circa* 15 mms. The rows of

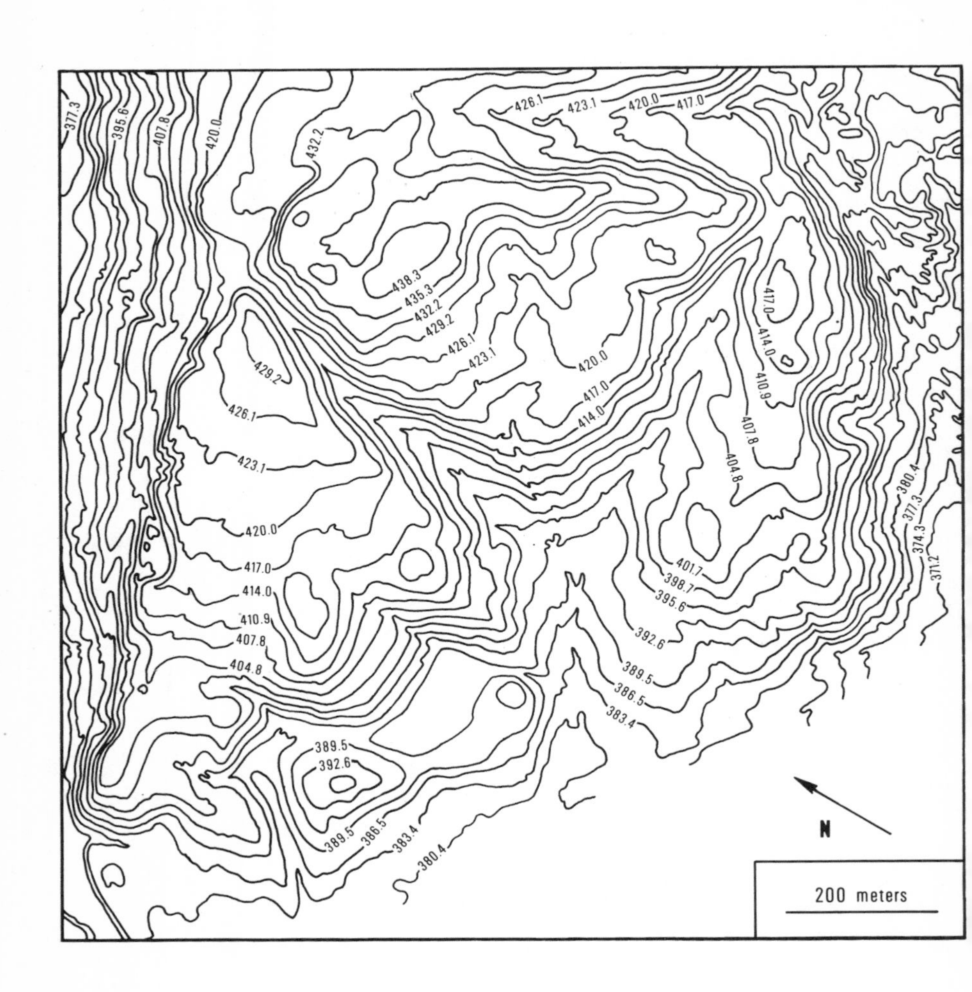
377.3
395.6
407.8
420.0
432.2
426.1
423.1
420.0
417.0
438.3
435.3
432.2
429.2
426.1
423.1
420.0
417.0
414.0
429.2
426.1
423.1
420.0
417.0
414.0
410.9
407.8
404.8
417.0
414.0
410.9
407.8
404.8
401.7
398.7
395.6
392.6
389.5
386.5
383.4
380.4
377.3
374.3
371.2
389.5
392.6
389.5
386.5
383.4
380.4
N
200 meters

Fig. 15.1.--Waunafon Opencast--Spoil contours

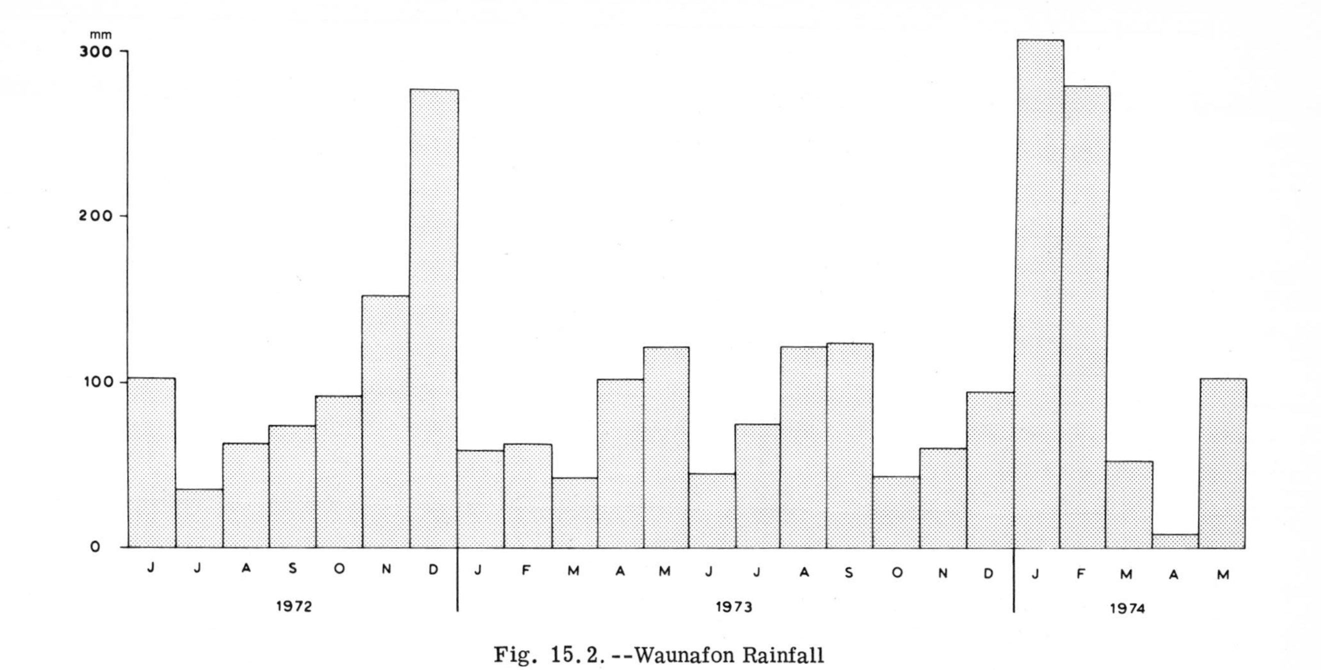

Fig. 15.2.--Waunafon Rainfall

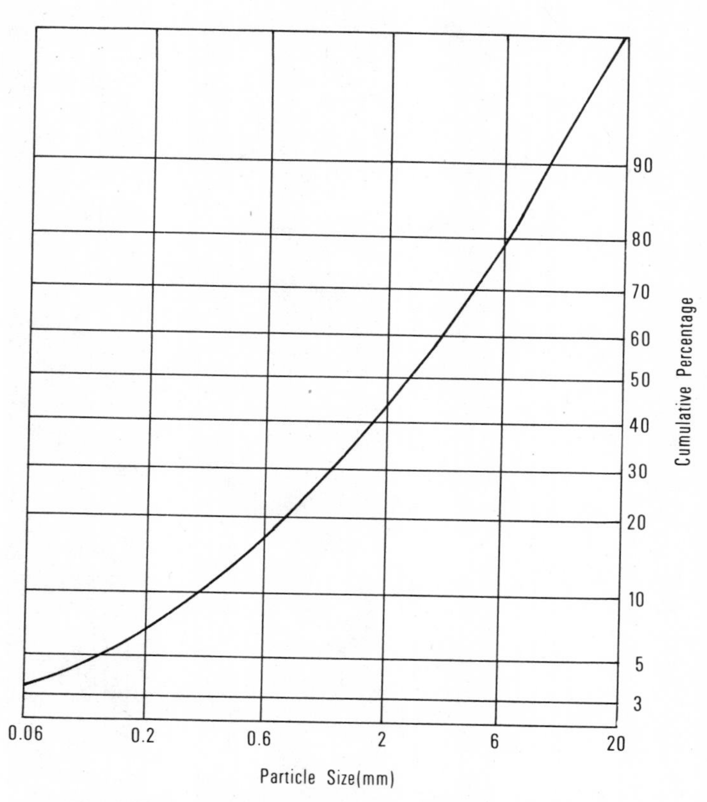

Fig. 15.3.--Particle Size Distribution, Waunafon

P. 15. 1. --The Reconstructed Valley at Cefn-Garn-yr-erw--Location of all of the Waunafon erosion grids

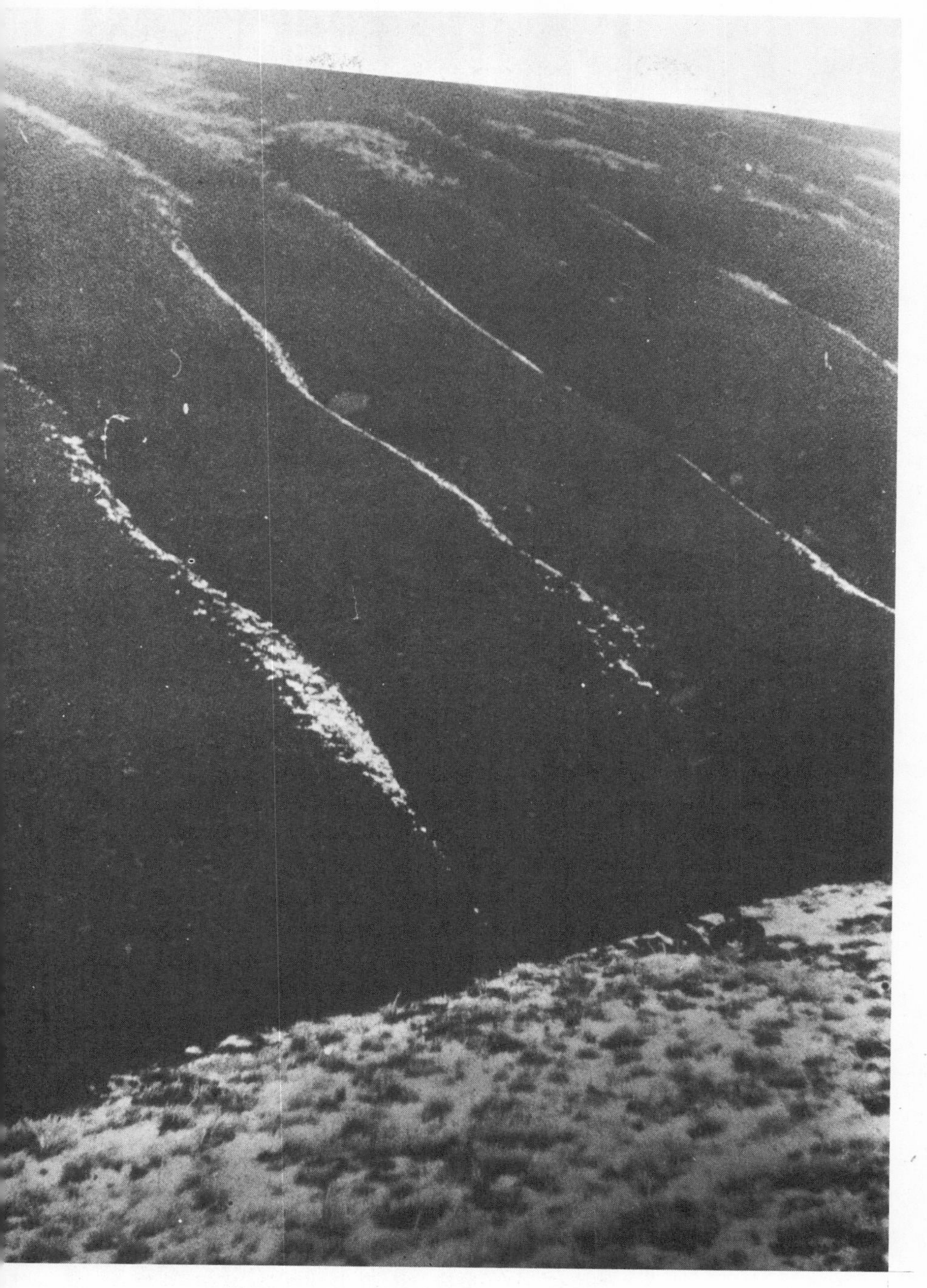

P. 15. 2. --Site of Waunafon D. Erosion Grid

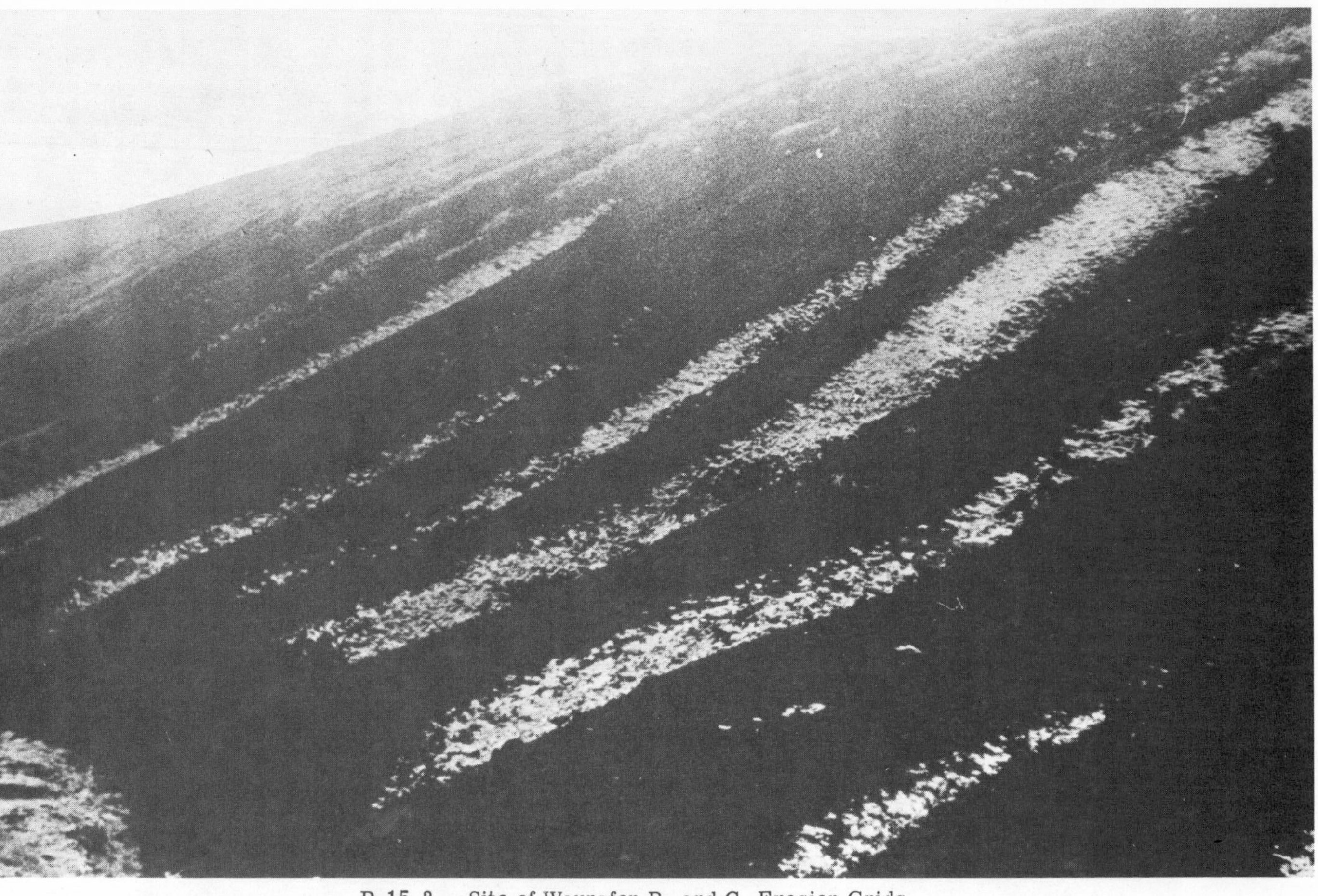

P. 15. 3. --Site of Waunafon B. and C. Erosion Grids

P. 15. 4. --Site of Waunafon W. Erosion Grid

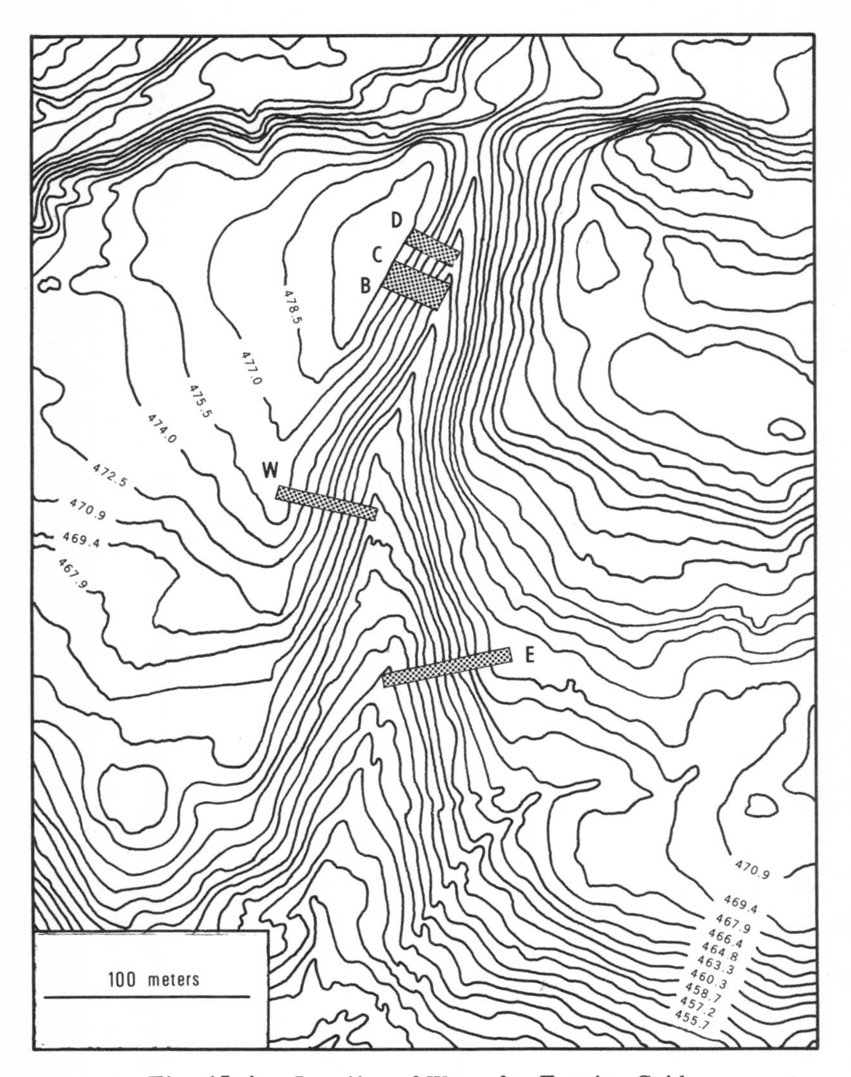

Fig. 15.4.--Location of Waunafon Erosion Grids

erosion pins are established at distances along the slope which are multiples of a 1.5-meter unit. The grid rows may be clustered in areas where there is a more rapid change in slope angle. Three of the experiments' erosion pins are fitted with close fitting permanent washers. These are Waunafon W., Waunafon E., and Waunafon D. Waunafon B., the gully profile, is recorded by means of a loose fitting removable washer which is placed over the pin immediately prior to data recording and removed immediately afterwards. Waunafon C. grid which is immediately adjacent to Waunafon B. and very near to Waunafon D. is a hybrid. Here, only the longer rods are fitted with permanent washers.

The experiments were all established during 1972. Waunafon B. and Waunafon C. grids were both established during April. Waunafon D. grid, and E. grid were added in late June while W. grid was not established until the first week in July. The first measurements were taken from B. grid and C. grid during June. The first measurements from the other grids were made immediately after their establishment. The grids have all operated successfully with a minimum of vandalism and disturbance throughout the course of the experiment.

Once again, it is probably best to start the consideration of the results from these experiments by a consideration of the "Quarterly ground retreat estimates" for each of the instrumented slopes. This function, which is intended to represent the mean vertical retreat of the whole profile, is calculated from the formula:

$$Q = 91/t \frac{(\Sigma [rw])}{\Sigma w}$$

where: Q is the quarterly ground retreat estimate;

t is the time in days between data records;

r is the per row difference in the exposure of erosion pins;

w is a weighting factor representing the distances separating each grid row from its neighbors; and the

"91," of course, is the number of whole days in a quarter of a year.

The cumulation of the quarterly ground retreat estimates recorded from the four instrumented slope profiles during the two years of the experiment is graphed as Figure 15.5. The results from the two vegetated grids are each represented by a single line on the graph. However, the results from the unvegetated grids are described by two lines each. The two unvegetated profiles, Waunafon C. and Waunafon D., are both characterized by a great deal of slope foot activity. The results from the gully channel and its immediate surrounds reflect relatively large changes as erosion is succeeded by massive deposition.

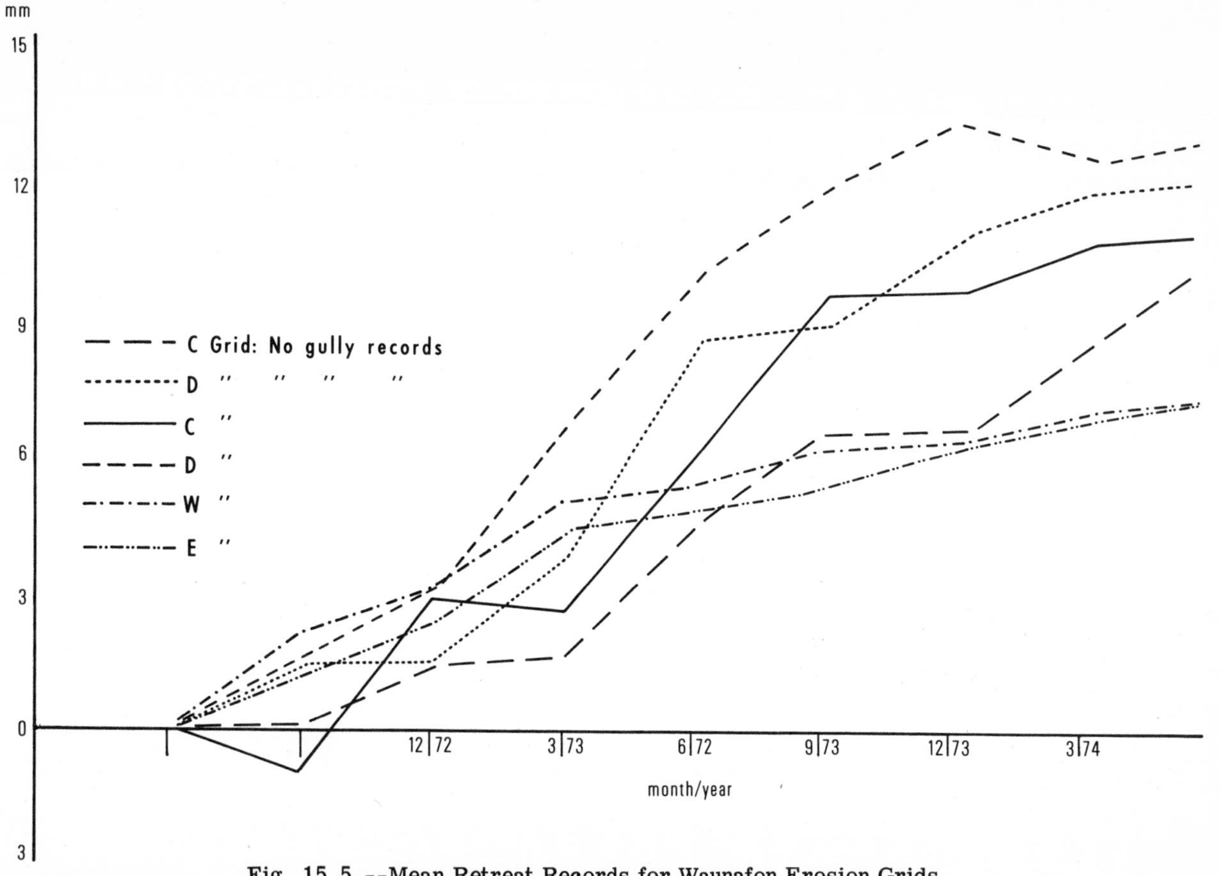

Fig. 15.5.--Mean Retreat Records for Waunafon Erosion Grids

It was felt that the inclusion of results which are obviously affected by the activities in the gully channel might tend to disguise, or reduce the credibility of, a quotient designed to summarize the erosion of the whole slope profile. Consequently, two additional lines have been added to Figure 15. 5 which represent the ground retreat recorded by the erosion pins which lie above the inflexion which marks the transition from slope proper to gully incision. Results from the lowest two rows of erosion pins, 0 meters and 1. 5 meters, which are situated in the gully channel incision, are excluded. It can be seen that the effect of the exclusion of these data records has reduced the apparent biennial retreat of the two profiles by between 0. 5 and 1. 0 mms.

The results in Figure 15. 5 suggest several conclusions. The first is that the unvegetated slopes retreat at a greater rate than do those which support vegetation. The four graphs from the vegetated slopes record a retreat of between 10. 5 and 13. 5 mms in the two years of the experiment. The two graphs from the vegetated slopes suggest that both slopes suffered a mean retreat of circa 7. 6 mms in the same period. The graphs from the two vegetated profiles are almost identical. It has been widely demonstrated that there are occasions when slopes of opposed aspect will suffer a very different degree of erosion. It has been found that this differentiation nearly always separates slopes of warm aspect, which face south south-west, from those of cold aspect, which face north north-east (vide: Geiger - 1965; Karrasch - 1968, 1972; Haigh - 1973, 1974b; Crabtree - 1972; Tinker - 1970). Differences tend to become reduced on slopes of intermediate aspect. The similarity of the results from the vegetated slopes of opposed aspect in Cefn Garn-yr-erw tends to support this assertion and might also be cited as testimony for the internal consistency of results gathered by this type of direct measurement.

However, there are certain important differences in the progress of mean retreat cumulation on the two slopes (Figure 15. 5). In all but two of the recording periods the west facing slope, Waunafon W., suffers a smaller rate of retreat than does the east facing slope, Waunafon E. The west facing grid, however, registers a vastly greater retreat during the two summers of the experiment. Waunafon W.'s mean retreat scores exceed those of Waunafon E. by 96% and 66% in the summers of 1972 and 1973 respectively. Waunafon E. registers its greatest relative retreat during the autumn of 1973 (596%) and in the two spring seasons (1972, 183%; 1973, 188%). These differences have a smaller overall importance, however, because the absolute retreat rates recorded in the spring seasons and in the autumn of 1973 were relatively tiny compared with the retreat scores of the summer months.

If one divides the two profiles at their mid-points and considers the two halves separately (cf. Gerlach - 1966) one discovers that the upper sections of Waunafon E. and Waunafon W., respectively, suffered only 42% and 70% of the retreat recorded on the lower slopes. Ground retreat on the lower slopes was exceeded by that recorded on the upper slopes on only 3 out of the 16 seasonal records. The probability of this situation not being indicative of a systematic trend could not be accepted at the p=10% level on either profile. However, it may be noted that the three instances of retreat on the upper slopes being greater than that on the lower slopes all occurred in the colder half of the year (Waunafon W., Winter 1973; Waunafon E., Autumn 1973, Winter 1974).

Table 15.1 is a summary of the results suggested by a comparison of the mean retreat scores on these two slopes. It can be seen that in both cases it is the upper slopes which suffer less retreat and that this tendency has been more marked on the west facing slope. One consequence of this tendency might be differential steepening.

TABLE 15.1

A COMPARISON OF THE RETREAT OF UPPER AND LOWER SLOPES

Grid	Cumulated Segregated Quarterly Retreat Estimates for the Period June 1972-June 1974		Ratio Upper/Lower Slopes
	Upper Slope	Lower Slope	Qu : Q1
Waunafon E.	5.2 mms	12.2 ms	1 : 1.4
Waunafon W.	6.2 mms	8.9 mms	1 : 2.4

Of course, the quarterly ground retreat estimate is too much of a generalization to permit much discussion of slope evolution. As has been implied by the duplication of graphs from the unvegetated profiles on Figure 15.5 and by the later independent considerations of the retreat of the upper and lower slopes, this quotient conceals a great deal of variation. An attempt has, therefore, been made to display this variation by the preparation of a series of diagrams in which the mean ground retreat at each row of erosion pins in the instrumentation grid is graphed against time.

The graphs in Figure 15.6 and Figure 15.7 are an attempt to display the

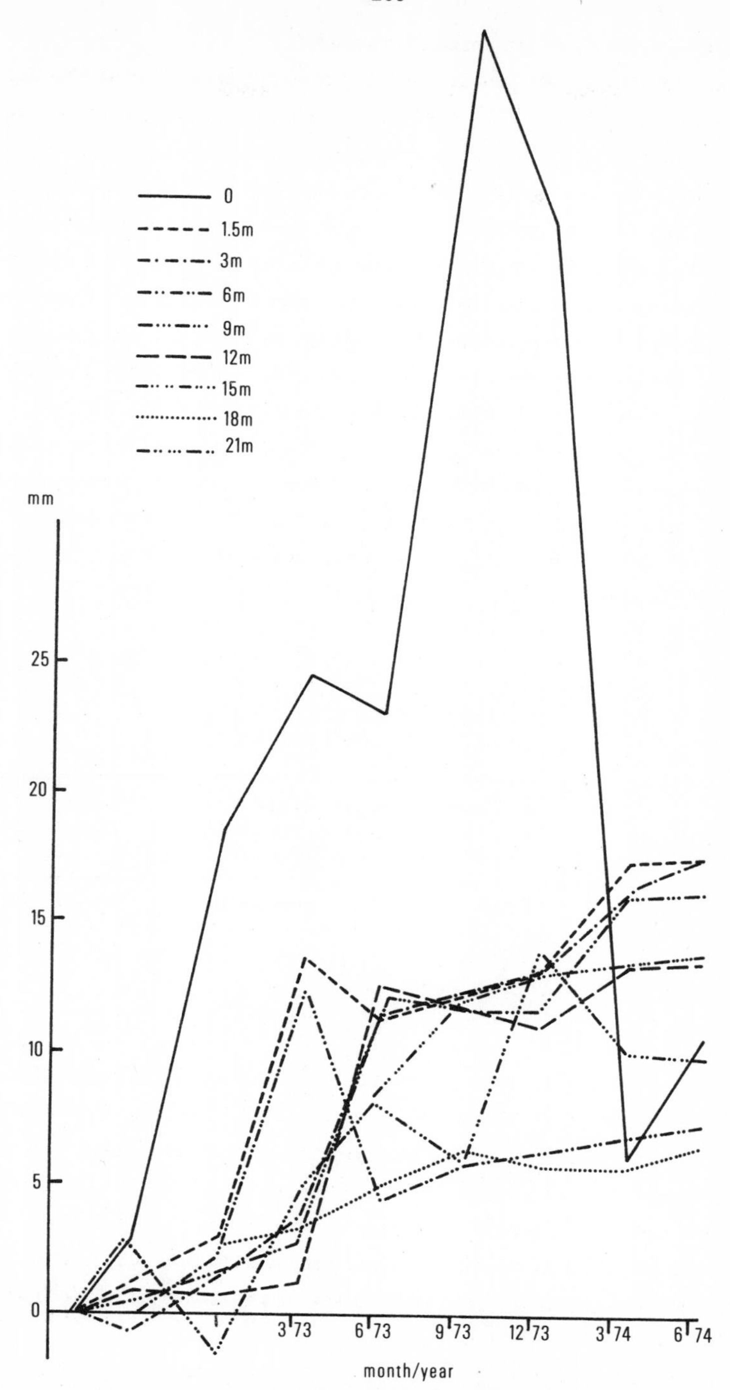

Fig. 15.6.--Waunafon D.: Ground retreat per grid row

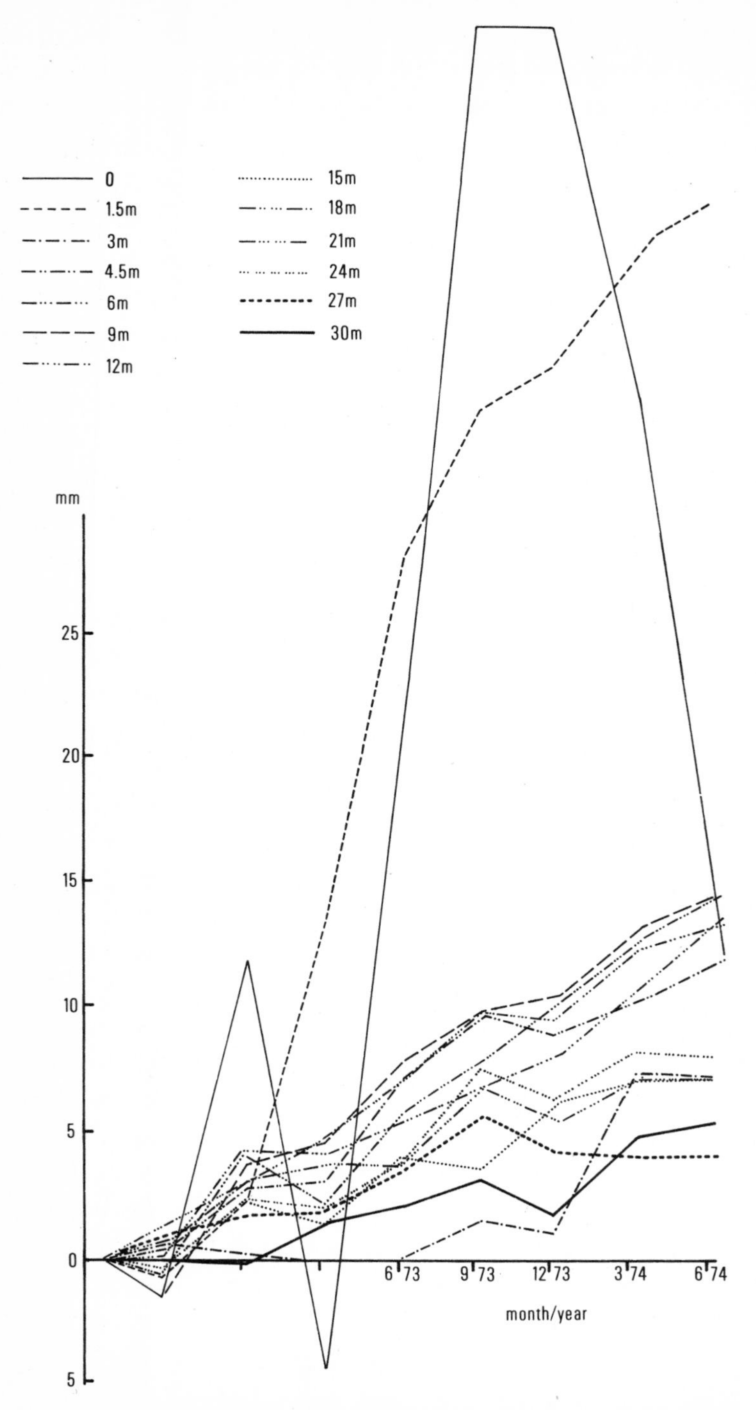

Fig. 15.7.--Waunafon C.: Ground retreat per grid row

variation in the ground retreat recorded from the different grid rows of Waunafon D. and Waunafon C., respectively. These two erosion grids are located very close to each other on the west facing slopes of Cefn Garn-yr-erw. Theoretically, the records of mean retreat from these two slopes (Figure 15.5) should have been as nearly identical as those from the vegetated grids which are separated by a greater physical distance and a different slope aspect. In fact, there is a considerable difference in the mean rate of erosion recorded on these two grids. It would be tempting to attempt to frame an explanation of this difference in terms of variations in the conditions of basal removal, and it is a fact that the removal of data from the gully channel zone erosion pins causes a slight reduction in the separation of the two slopes' mean retreat rates (Figure 15.5). However, Figure 15.6 and Figure 15.7 demonstrate that the slopes of most of the graphs from most grid rows of both experiments are very much the same. In fact, there would seem to be just two important differences between the results from the two profiles.

The first difference is that during the first half of 1973 there was a marked increase in the rates of retreat affecting all but the lowest slopes of Waunafon D. (Figure 15.6 Grid rows between 3 and 21 meters). This acceleration of erosion was imitated but not equalled by the records from the middle and upperslopes of Waunafon C. (Figure 15.7 Grid rows between 6 and 24 meters). Now, it is possible that some of the acceleration of retreat registered on Waunafon D. might be due to the frost-heave disturbance of the erosion pins. The record from the 6-meter grid row during the Autumn of 1972 is a good example of the effect of frost-heave. The results from this row register a massive increase in retreat as the pins are raised from the ground. This is counterbalanced by an equally massive ground advance as the ground thaws and the pins settle back into, more or less, their old positions. Frequently, such disturbances affect only part of a grid row and, usually, can be edited from the data record. It is possible that the irregular oscillations of the graph from the 15-meter grid row also indicate that this area was subject to some form of intermittent disturbance. However, it is not plausible to attribute the acceleration of erosion which affects all the mid-slope profiles, to frost action since this acceleration occurs concurrently with the closely observed recovery of the pins of the 6-meter grid row from the frost-heave of the previous season.

There is a second major difference between the two diagrams. This arises from the behavior of the 1.5-meter grid row on Waunafon C. (Figure 15.7) which is very different from that on Waunafon D. (Figure 15.6). This trace, on both slopes, is that which records the greatest retreat. However, the 1.5-meter trace on Waunafon C. registers nearly twice the amount of retreat of that on Waunafon D. Further, the 3-meter trace which, on Waunafon D. records just as much retreat as the 1.5-meter trace, records only about a sixth of the 1.5-meter retreat on Waunafon C. Thus, while the 3-meter grid row on Waunafon D. was located at a site which suffered a notably rapid rate of retreat during both years of the experiment, the three-meter grid row on Waunafon C. occupied a site which was neutral during 1972-1973. The period during which the 3-meter trace on Waunafon C. remains neutral coincides with that when the 1.5-meter trace was undergoing a spectacularly rapid retreat. During the second year of the experiment the slopes of the graphs from both grid rows are closer to the norm for these two erosion grids. One may conclude, therefore, that the differences in the erosion grids' scores are probably not a result of differences in the conditions of basal erosion, but of differences in the way basal erosion affected the two slope profiles. On Waunafon D. the effects of the basal channel were transmitted upslope to influence erosion on all the lower slope. On the longer slope, Waunafon C., the influences of the gully channel extended upslope only to the 1.5-meter grid row, at least during the first year of the experiment.

There is a third factor which may have contributed slightly to the creation of a difference in the quarterly retreat estimates from the two grids. This is that the grid on Waunafon D. was not established as successfully as that at Waunafon C. A result of this has been that there are fewer results from the upper element of the smaller slope. It may be seen that the retreat scores of the upper slopes of Waunafon C. (Figure 15.7: Grid Rows at 27 and 30 meters) are among the smallest on this profile--not least because these areas still retain vestiges of their original vegetation. This fact alone, however, would not be sufficient to account for the degree of separation apparent in the mean slope retreat estimates shown in Figure 15.5.

Analysis of the ground retreat record suggested by the data graphed in Figure 15.6 and Figure 15.7 is often made easier if the data are presented in another way. In Figure 15.8 and Figure 15.9 the data from each three-month data period recorded during the experimental period are graphed against the distance upslope. The first diagram, Figure 15.8, represents the results from Waunafon D. with the exception of data from the gully channel. Here, the results from the four seasons of the two years are graphed side by side. It can be seen that there is little correlation in results from season to season and that,

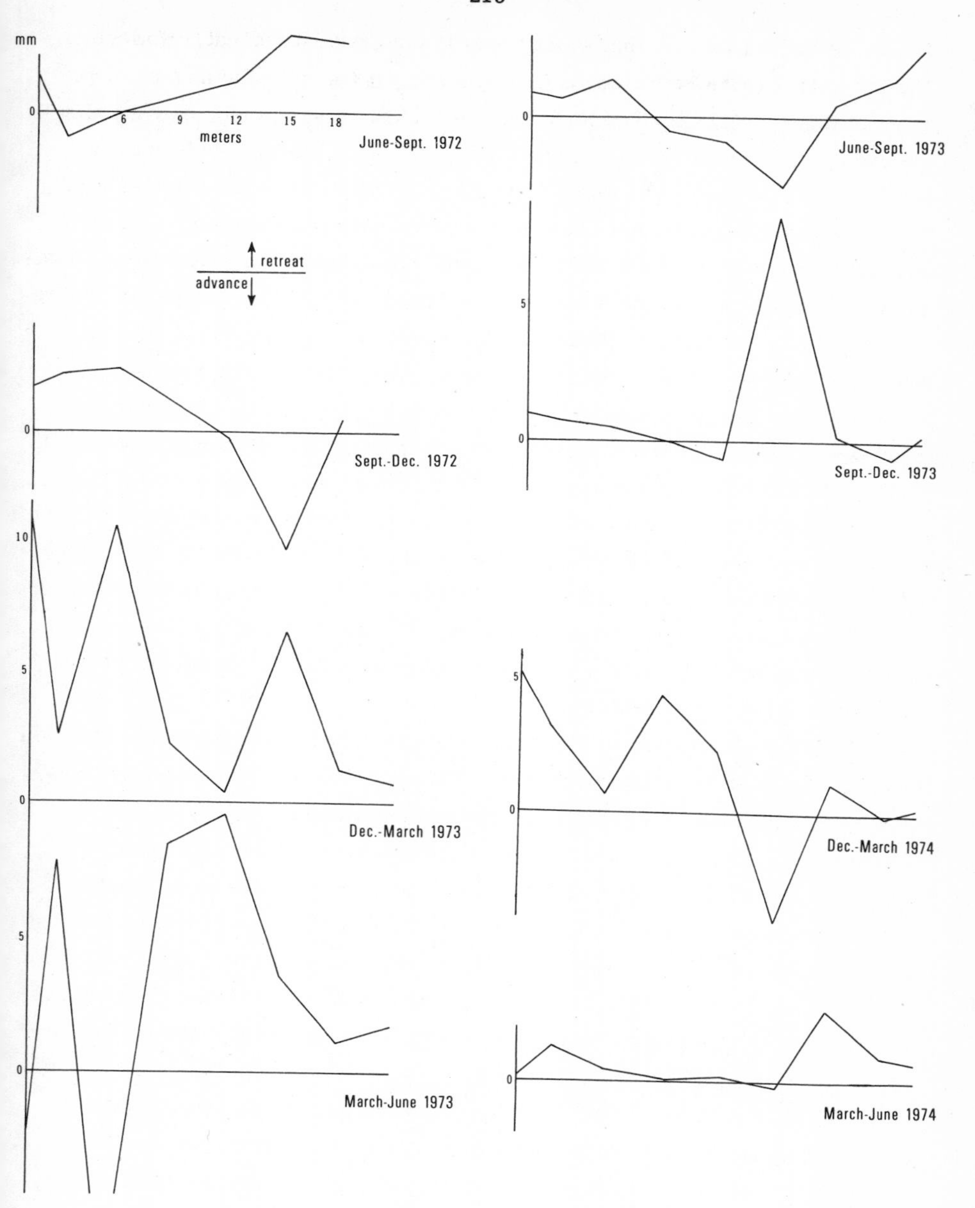

Fig. 15. 8. --Waunafon D.: Upslope variations in ground retreat for each quarter year.

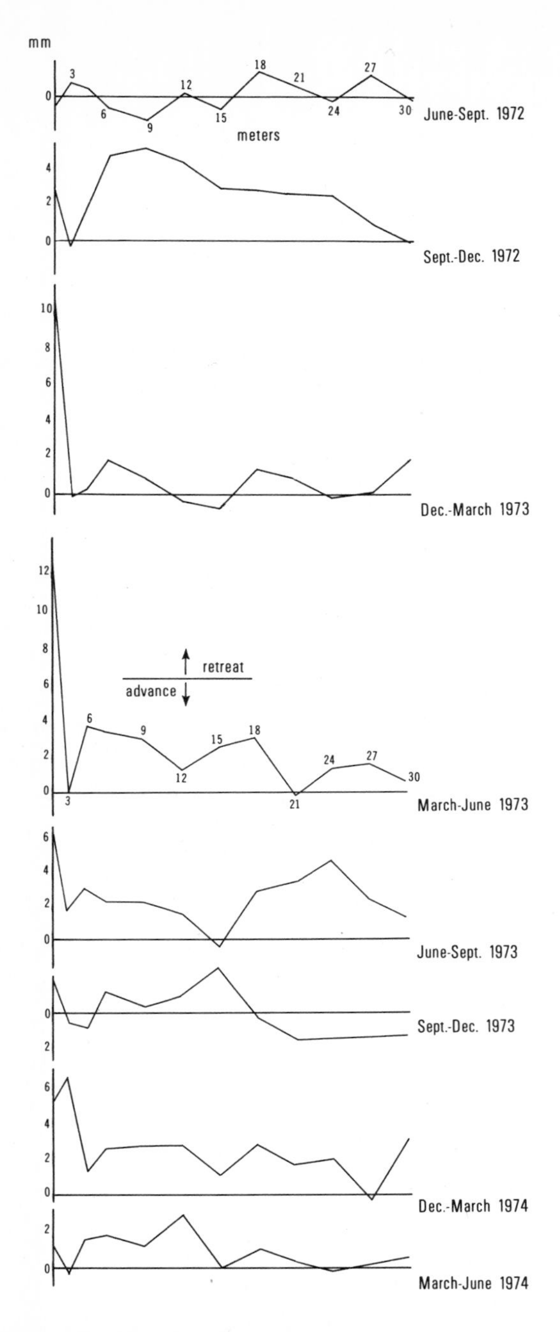

Fig. 15.9.--Waunafon C.: Upslope variations in ground retreat for each quarter year.

in fact, these graphs are best treated as a linear series. Figure 15.8 emphasizes the irregularity of the data from Waunafon D. and highlights again the strange behavior of the 15-meter grid row's data records.

The graphs from Waunafon C., presented as Figure 15.9, present a far more reasonable picture. This diagram, again, excludes data from the gully channel because of the problem of scale. The slow acceleration of the retreat rate recorded by the 1.5-meter grid row is, however, clearly demonstrated by the first four graphs, while the next three are suggestive of the transmission of this stimulus to the 3-meter slope zone. The retreat scores from the remainder of the slope re-emphasize the fact that it is the lower and mid-slope areas which tend to record higher rates of ground retreat. Incidences of neutrality and ground advance tend to be restricted to the upper slopes and the 3-meter slope area--which is the slope foot adjacent to the gully banks. More controversial might be the suggestion that it is possible to detect the upslope propagation of waves of a particular erosional activity--for example the ground advance phase affecting slope at 15 meters in the period June-September 1973; the whole slope above 18 meters in the following autumn; and the slope at 27 meters in the winter of 1973-1974 (cf: Marosi - 1972; Pinczes - 1972).

Figure 15.10 and Figure 15.11 illustrate the cumulation of the ground retreat scores recorded on the individual erosion grid rows of the two vegetated grids Waunafon W. and Waunafon E. The results from Waunafon W. are especially interesting because this is the profile which has the least effective basal gully control. In fact, there has been no gully incision at all at the foot of this slope since the site was restored. It is significant thus that, on this slope alone, the results from the slope foot are not essentially different from those on other sections of the slope. The erosion pins which record the greatest retreat on this slope belong to the rows of the lower mid-slope (Figure 15.10 Grid rows at 12 and 15 meters). The situation on Waunafon E., is rather more familiar. Here, there has been a degree of basal erosion and this is reflected by erosion records from the slope foot (Figure 15.11 Grid rows at 0 and 1.5 meters).

The results at Waunafon W. (Figure 15.10) also possess one feature in common with those from the unvegetated Waunafon C. (Figure 15.7). This is the relative inactivity of the 3-meter grid row. There is no 3-meter grid row on Waunafon E. but, here, the 6-meter row, which in terms of slope morphology is situated in a position which is analogous to that of the 3-meter rows of the other grids, is also that grid row which records the least ground retreat.

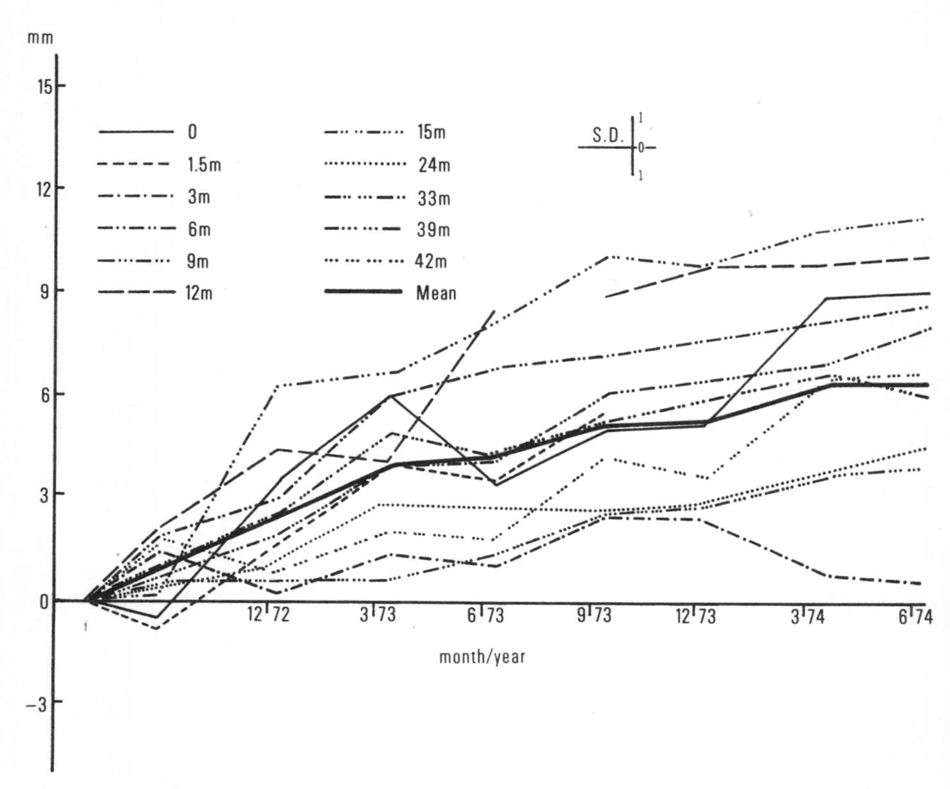

Fig. 15.10.--Waunafon W.: Ground retreat per grid row

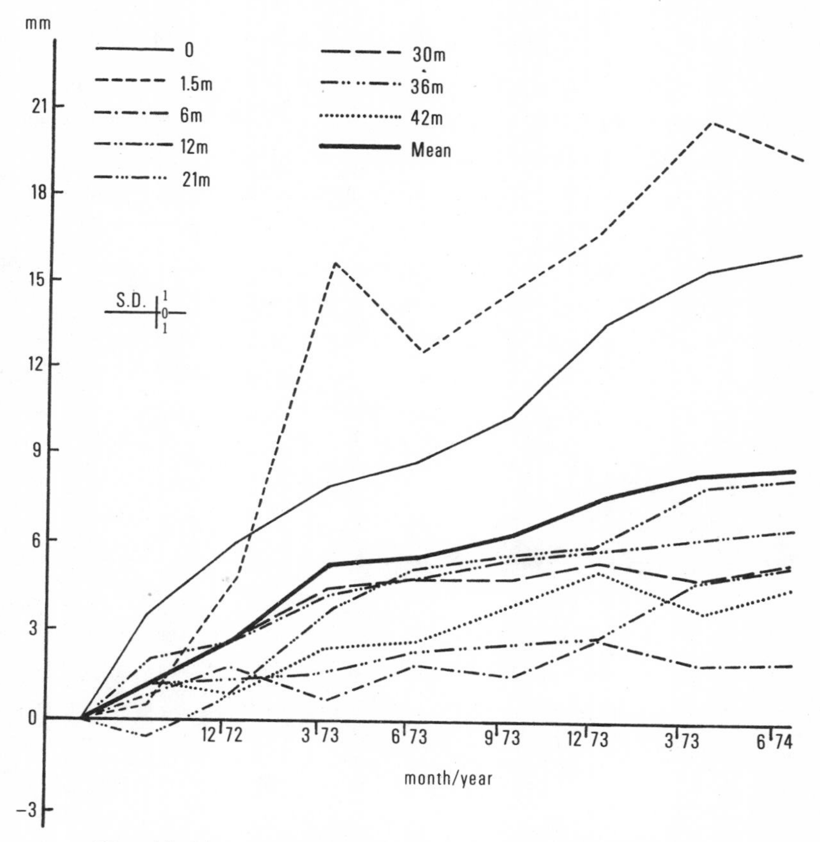

Fig. 15.11.--Waunafon E.: Ground retreat per grid row

The variations of retreat detected during each individual recording period have been graphed against distance upslope for both the vegetated profiles in Figure 15.12 and Figure 15.13. It can be seen that there is little obvious correlation of activity either by season on individual profiles or between profiles in individual seasons. A comparison of the two graphs emphasizes that even on the profile with the less active basal control, ground retreat in the channel section of the lower slope is substantially independent from activity on the rest of the profile (Figure 15.12). The results from the slope with the more active basal control, Waunafon E. (Figure 15.13) show essentially the same thing, though the possibility of the transmission of retreat stimuli upslope might, here, be admitted. On both slopes, maximum erosion rates on the non-gully channel slope sections are concentrated in the lower mid-slope (Waunafon W., Figure 15.10 and Figure 15.12, Grid rows at 12 and 15 meters; Waunafon E., Figure 15.11 and Figure 15.13, Grid rows at 12 and 21 meters).

It is possible to gain a more precise picture of the implications of these two years' data for the process of slope evolution by representing the information in another way. A third series of graphs has, thus, been prepared in which average annual ground retreat is recorded on a representation of the actual slope profile. These graphs present the process of slope evolution in its simplest guise, but just as was the case with the graphs of the quarterly ground retreat estimate, their construction involves quite a high degree of generalization.

The index of retreat employed by the third series of graphs is the average annual retreat registered at each grid row. As the preceding analyses demonstrate, retreat varies irregularly from recording period to recording period and from station to station across the whole slope. There is little detectable seasonality, at least when the data are analyzed at grid row level. However, there is a major difference in the amounts of retreat recorded in the two years of the experiment. The unvegetated slopes record 60% of their total retreat, and the vegetated grids 70% of their total retreat, in the first year of the experiment.

Further, the slope profile on which these results are represented is not a true slope profile. This profile is a construct developed from the concept of flow line slopes. Measurements were made of the flow line slope angles over a meter-unit slope above and below each individual erosion pin. The mean upslope and down-slope flow line slope angles for each grid row were then calculated and graphed about the height/distance co-ordinates of the grid row in ques-

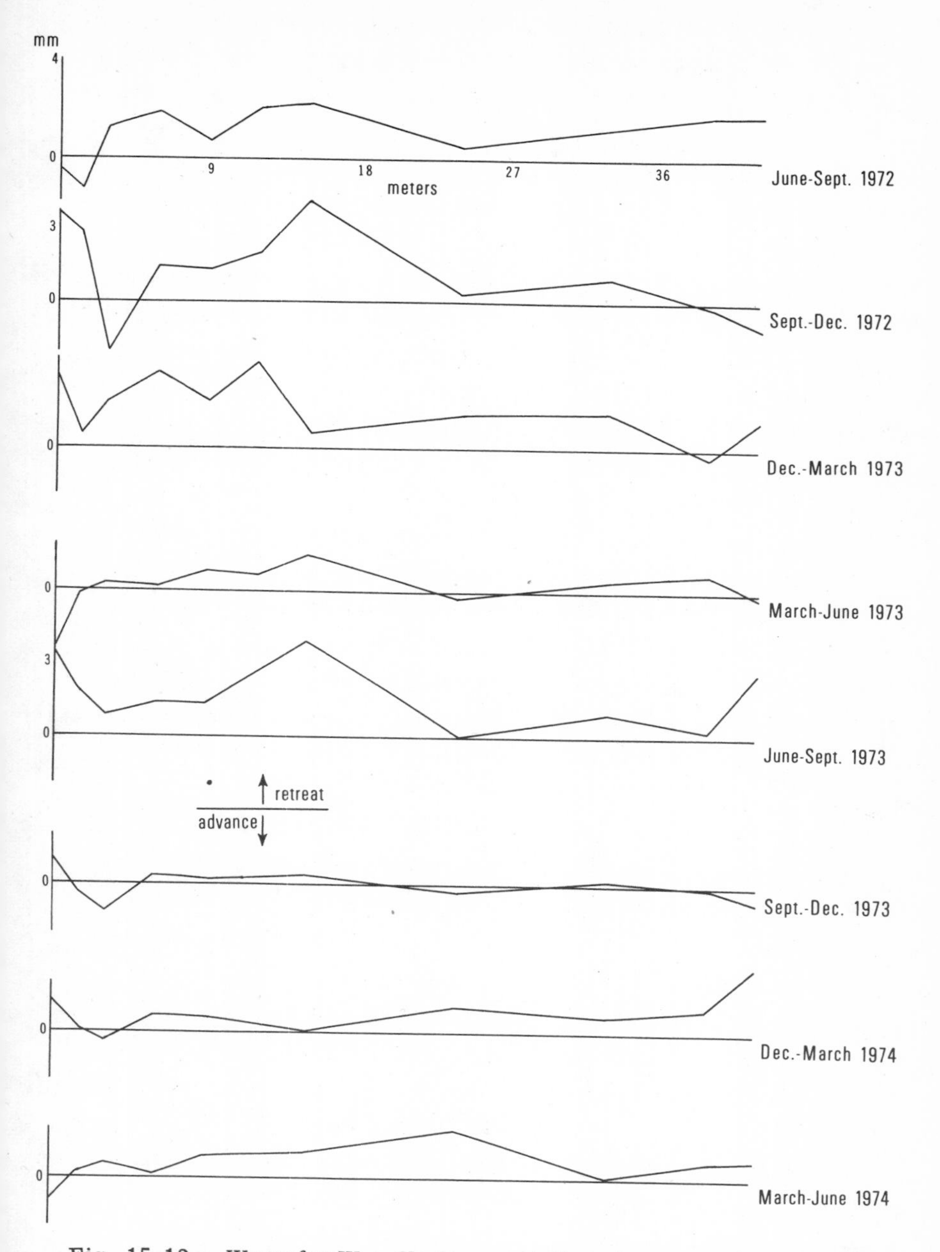

Fig. 15.12.--Waunafon W.: Upslope variations in ground retreat records for each quarter year.

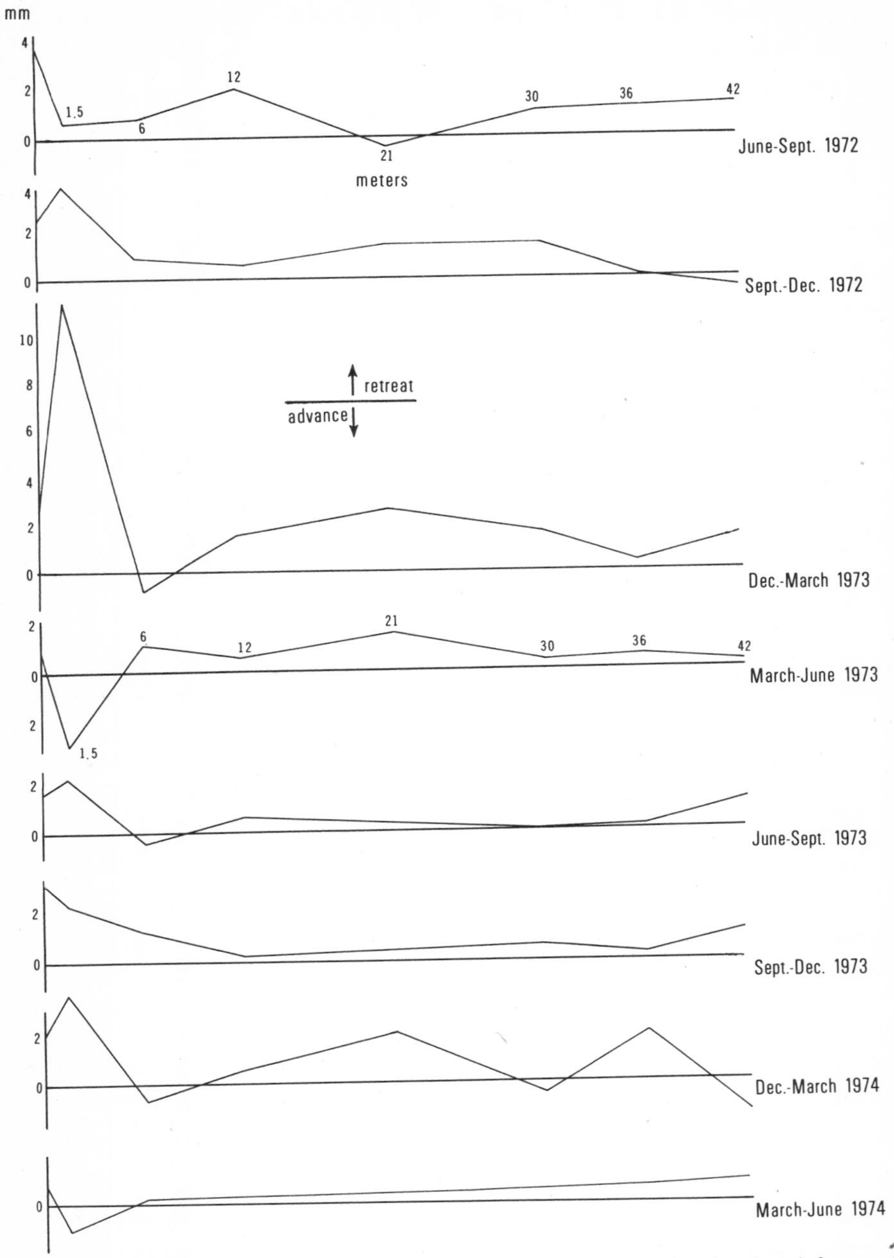

Fig. 15.13.--Waunafon E.: Upslope variations in ground retreat for each quarter year.

tion, and the graphed lines from each grid row extended until they intersected. It was felt that the slope profile thus generalized was a better indication of the instrumented slope than was the ordinary meter-unit profile because it ignored morphological detail whose effect on slope evolution was not monitored and could not be assessed.

Naturally, recorded ground retreat is not of the same order of magnitude as the gross slope profile. The ground retreat results have had to be magnified by a factor of 100 to achieve the visual impact of these graphs. The visual effect of this multiplication might be regarded as a representation of the assertion that, if the average annual retreat record from the two years 1972/3-1973/4 is similar to the average annual retreat record of the period 1972-2072, then these graphs represent the change of slope morphology which will occur in that period. Table 15.2 catalogues the differences in the amounts of retreat suffered by the four grids in the two years of observation. These differences were also reflected in the distributions of erosional activity on the slope profile. The following discussions, thus, should be treated with caution.

TABLE 15.2

ANNUAL RETREAT RATES--WAUNAFON

Grid	June-June		Annual Average
	1972-1973	1973-1974	
Waunafon C.	6.25	4.76	5.50
C (no gully)	4.55	5.02	4.78
Waunafon D.	10.20	2.84	6.52
D (no gully)	8.60	3.65	6.12
Waunafon W.	5.46	2.04	3.75
Waunafon E.	4.99 mm	2.55 mm	3.77 mm

Figure 15.14 represents the results from the two unvegetated profiles: Waunafon C. and Waunafon D. The results from Waunafon C. are dominated by morphometric alteration in the gully channel area and at the junction between the main slope segment and the upper slope element. The results from the slope foot suggest that undercutting is causing massive parallel retreat of the gully bank which is a consequence of the increasing depth of gully incision and the extension of the gully bank segment into the lower slope element. This slope unit is developing an increased curvature and thus becoming more distinctly separated from the upslope main slope segment. Retreat of the main

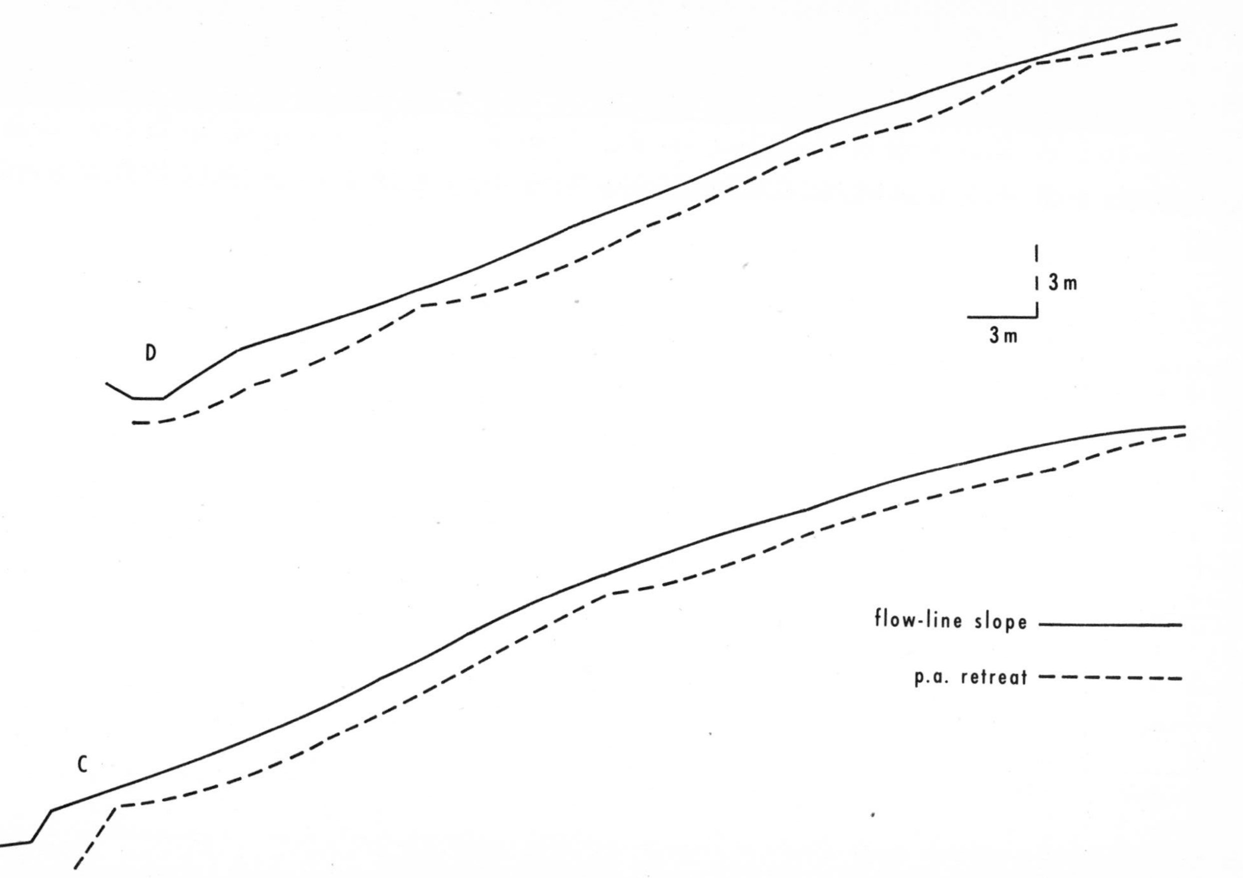

Fig. 15.14.--Retreat x 100 Graphed on Flow Line Profiles of the Unvegetated Slopes: Waunafon C. and D. grids

slope segment is characterized by parallel retreat and abbreviation due to the extension of the upper and lower slope elements. The works of Young and Mutchler (1967; 1969a, b) and Meyer and Kramer (1968, 1969) all support the contention that an independent convex slope will tend to evolve towards a concave profile. This assertion is supported by an examination of retreat recorded on the upper convex element of Waunafon C. It is, however, uncertain whether this process would lead to a hollowing at the upper convexities' junction with the main slope or whether this effect would migrate along the profile to encourage a more rapid extension of this slope unit and, possibly, the internal development of a new upper slope segment. Evolution at the crest is characterized by slope decline and an increased abruptness of the junction between the crest segment and the upper convexity. These projected morphometric differentiations are, of course, accomplished by a relatively small proportion of the total retreat and occur within a picture which might be crudely generalized as uniform ground loss across the whole slope (cf. Young - 1972; Schumm - 1956).

The pattern of slope evolution suggested by the results from Waunafon D. is essentially the same. One may recognize decline of the crest segment, the hollowing out of the upper convexity, the parallel retreat of the abbreviated main slope segment, and the increasing differentiation of a lower slope element. This effect, however, is disguised by a very much more rapid retreat of the break separating the slope foot from the gully banks, and this disguise is compounded by the fact that these gully banks, apparently, do not retreat parallel but actively decline.

The results from the two vegetated profiles are very different, except in the vicinity of the gully channel (Figure 15.15). There are, again, massive retreat scores from the gully banks which effect either a positive or negative decline of this segment. There is, also, an implication of the persistence of a basal slope element of possibly increased curvature and the maintenance of an abrupt junction to separate this from the gully banks. The graph from Waunafon W. contains a suggestion of slope steepening by the creation of a concave element in the lower main slope section. This process is less apparent in the graph from Waunafon E. where parallel retreat dominates the whole profile except the slope foot and the upper element--which register a small increase in curvature. It seems that slope evolution on both profiles is characterized by the parallel retreat of all slope units except those affected by the retreat of the gully banks and except for the lower main slope segment of Waunafon W. which is indicative of a slight tendency towards the negative slope decline of that profile.

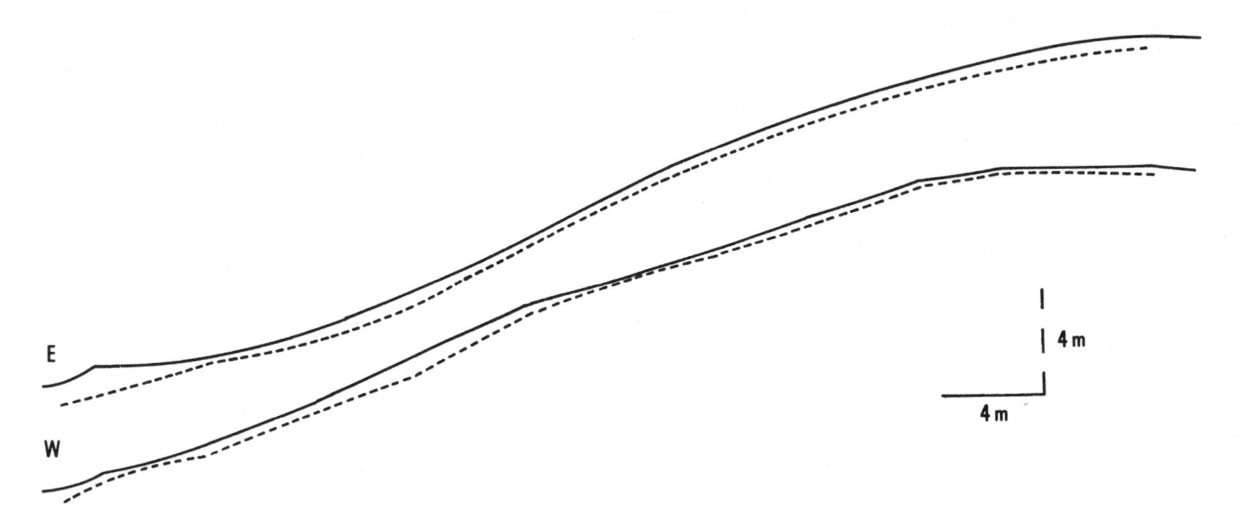

Fig. 15.15.--Retreat x 100 Graphed on Flow Line Profiles of the Vegetated Slopes: Waunafon E. and W. grids

Finally, it is worth considering the behavior of the instrumented gully channel in the light of the observations concerning interfluve slope development. The results from the gully channel are briefly summarized as Figure 15.16. This diagram follows the pattern established by Figure 15.14 and Figure 15.15. Gully incision is graphed against, in this case, the actual slope profile and the original gully incision. Rates of gully erosion are much higher than those of normal slope retreat. It was necessary to multiply the annual average retreat only by a factor of twenty to magnify the gully data to a scale appropriate for this diagram. It can be seen that the evolution of the gully long-profile echoes the development of the original slope. The measured incision amplifies the trends established in the incision already achieved. The retreat data records a progressive increase in the rate of incision down slope from the gully head. This rate of increase becomes magnified at the junction of the upper slope element with the main slope segment and maximum incision is achieved at the base of the main slope. Below, there is an abrupt reduction in incision rate which seems related to the existence of the basal slope element. The occurrence of a zone of relatively neutral slope retreat separating the main slope and gully bank is a feature of all of the Waunafon profiles. Beyond this zone, the rate of incision accelerates towards the gully channel. The pattern is clearly displayed on the graph of net incision contained by the same diagram (Figure 15.16).

To conclude: ground retreat and slope development have been monitored for two years on four slope profiles and in one gully channel developing on the fill of the former opencast mine at Waunafon, in the artificial valley at Cefn Garn-yr-erw. It is demonstrated that ground retreat is far more rapid on the profiles which lack vegetation than it is on those which are vegetated. The results from the sections of the two unvegetated profiles which are not contained by the basal gully incision are 6.12 mms p.a. (Waunafon D.) and 4.78 mms p.a. (Waunafon C.). The differences are a result of variations in the upslope transmission of the influences of basal gully erosion. The second grid includes a greater number of erosion pins and, probably, is a slightly better estimation of the usual retreat of such slopes. The two vegetated profiles are sited on slopes of opposed aspect and experience slight differences in their conditions of basal removal. In spite of this they undergo an almost identical amount of ground retreat. The mean retreat of the east facing slope, Waunafon E., is calculated to be 3.77 mms p.a., and that of the west facing slope, Waunafon W., 3.75 mms p.a.

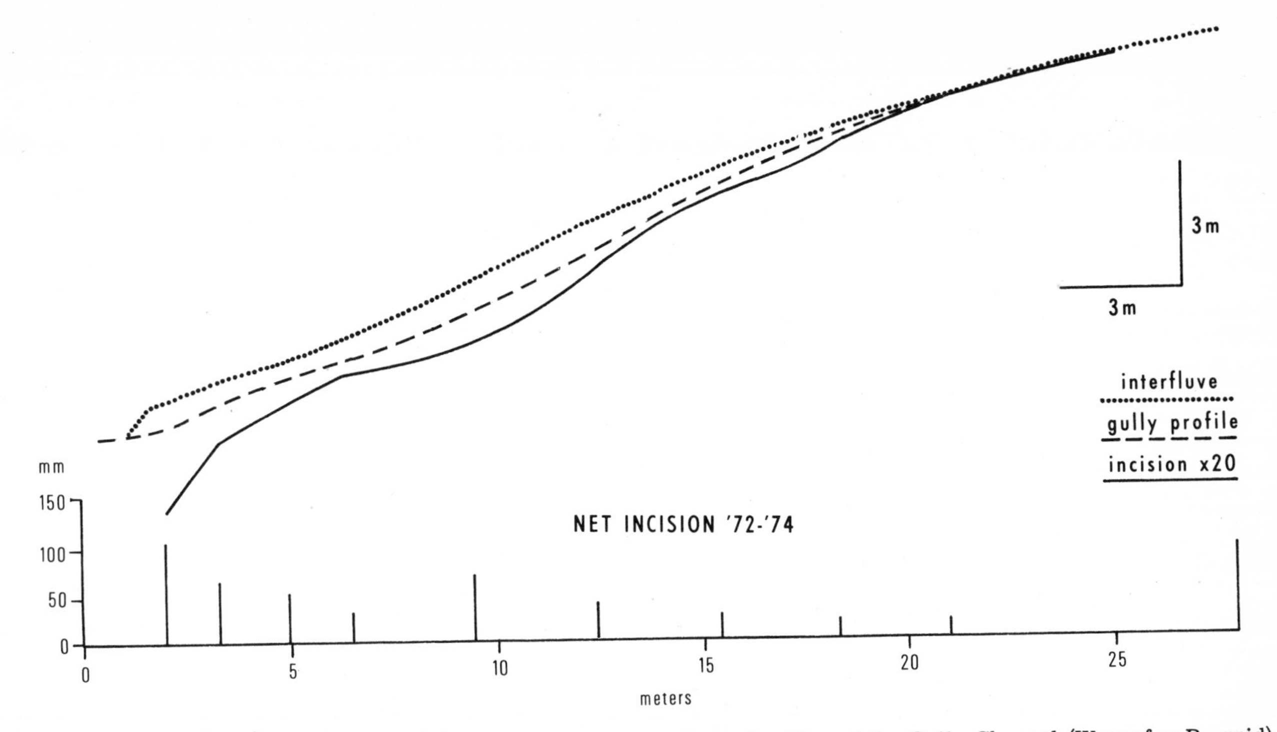

Fig. 15.16.--Gully Retreat x 20 Graphed against Flow Line Slope Profiles of the Gully Channel (Waunafon B. grid) and Adjacent Unvegetated Slopes (Waunafon C. grid).

P. 15. 5. --Waunafon Opencast Fill Viewed from Llanelly Hill

There are certain differences in the amounts of retreat affected on the two types of slope in the two years of observation. Retreat during the first twelve-months of the experiment, June 1972 to June 1973, exceeded that recorded in the following twelve months by an average of 65%: 70% on the vegetated profiles, 60% on the unvegetated profiles.

Slope evolution on all four profiles was mainly effected as parallel retreat except in the vicinity of the gully channel. This was especially true for the vegetated slopes of opposed, but ostensibly "neutral," aspect. The aspect difference encouraged a slight tendency towards differential slope steepening effective on the west facing profile. However, this tendency was a more impor tant influence on these profiles than differences in the activities of their both rather quiescent basal controls. The results from the unvegetated profiles suggested a more complicated pattern of slope evolution which was dominated by rapid gully bank recession, the differentiation of a well developed basal slope element above the gully incision, the parallel retreat and abbreviation of the main slope segment, the extension of the upper element and the decline of the crest segment.

CHAPTER 16

GROUND RETREAT AND GULLY DEVELOPMENT ON THE VEGETATED AND UNVEGETATED SLOPES OF THE INFILLED OPENCAST SITE AT WAUN HOSCYN

The Waun Hoscyn Opencast was the third of Blaenavon's open pit mines. It was excavated on a wide valley bench, high on the western slopes of the Afon Lwyd Valley. Its spoil banks stretch some 3 km from Forgeside, Blaenavon (N. G. Ref: 244082) in the north to Varteg (N. G. Ref: 254055) in the south, and are backed up against the valley wall from an altitude of around 380 meters to almost 470 meters.

The mine was excavated in the later 1940s by the same American machinery which was used for the war-time opencast operations to the north of Blaenavon. The site was infilled and restored to rough mountain grazing by its coaling contractor in 1956. Unfortunately, this attempted restoration was not successful. Vegetation regrowth was patchy and too thin to prevent a major phase of gully dissection. By the late 1960s, the site had reverted to a "badlands" type appearance (Photograph P. 16. 1) which was indistinguishable from that developing on Waun Hoscyn's unreclaimed sister sites at Waunafon and Pwll Du.

This unsatisfactory situation was recognized by the Opencast Executive of the National Coal Board which had inherited responsibility for all of Blaenavon's derelict opencast workings. A new phase of reclamation was initiated in June 1970. The Opencast Executive sponsored, on a competitive basis, an experimental seeding of eleven 30-meter strips laid out on the site ". . . as a means of evaluating hydromatic and other patent methods of seeding with conventional methods of seeding . . ." (N. C. B. Opencast Executive - 1972).

The aims of this study are less complex: simply, to discover the effect of a dense seeded turf on the rates of ground retreat on interfluve slopes and in associated gully channels. Two slope profiles are being monitored at Waun Hoscyn (Figure 16. 1), both are located on the same, northeast-facing slope, they have a similar inclination, and similar profile configurations.

The cardinal difference between the two instrumented profiles is the dif-

P. 16.1. --Gullying Adjacent to Waun Hoscyn U. Grid

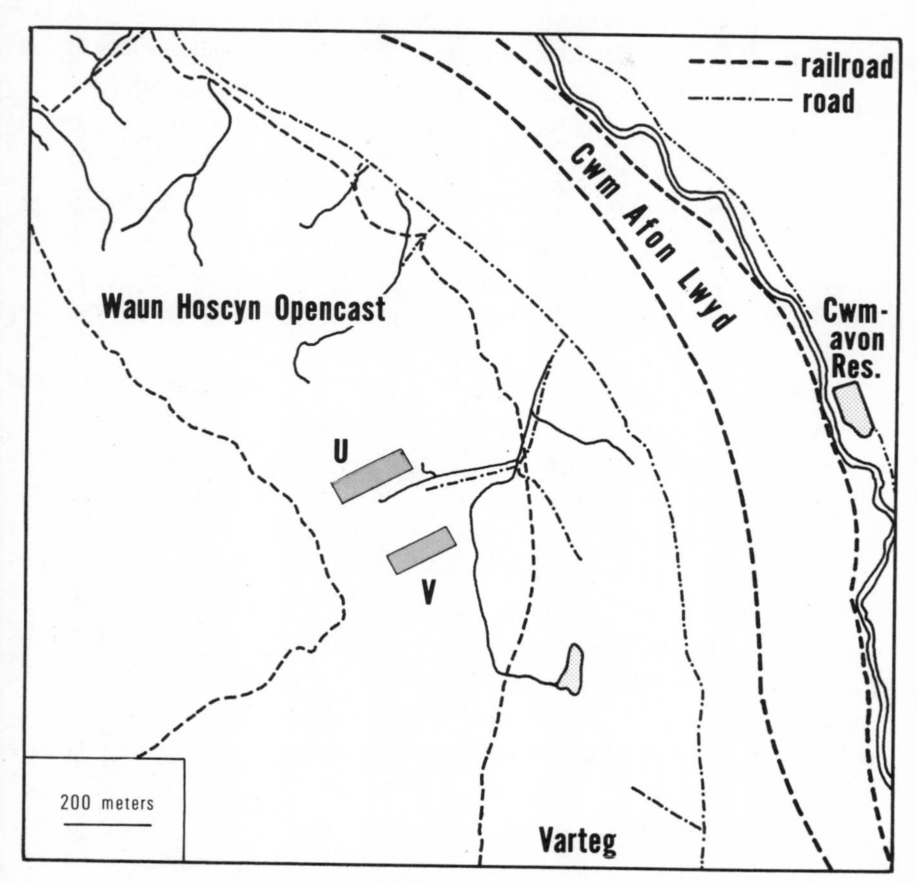

Fig. 16.1.--Location of Waun Hoscyn Erosion Grids

ference in vegetation. Waun Hoscyn "V." (Photograph P. 16. 2), is covered by a dense sheep-cropped turf. Waun Hoscyn "U.," less than 100 meters distant, supports a very poor vegetation consisting mainly of occasional grass tufts.

The main species on Hoscyn "V." are: Agrostis stolonifera L., Molinia caerulea (L.) Moench., Festuca ovina agg., Trifolium pratense L., Luzula campestris (L.) DC., and Trifolium micranthum Viv., together with ruderals such as Taraxacum officinale Weber, Cirsium arvense (L.) Scop., C. vulgare (Savi) Ten., Cerastium spp., Sonchus spp., Plantago spp., also Potentilla erecta (L.) Raüschel, Carex spp., and the grasses Lolium perenne L., Anthoxanthum odoratum L., and Cynosurus cristatus L. These flowering plants are supplemented by several species of moss and lichen. Vegetation cover is always complete, except on the banks of gully channels where it may be as little as 50% in winter.

The vegetative cover of Hoscyn U. is far thinner and contains far fewer species. Cover rarely exceeds 50% and is characteristically about 30%. The most common species is Festuca ovina agg., which is supplemented by tufts of Agrostis tenuis Sibth., Nardus stricta L., and locally, Juncus effusus L. Several species of moss, lichen, and algae are well represented.

Climatically, the slopes studied at Waunafon and Waun Hoscyn are very similar. There is no direct record of rainfall at Waun Hoscyn itself, but the rainguage at Cwmavon Reservoir, situated at 258 meters on the floor of the deeply incised and narrow Afon Lwyd Valley (N. G. Ref: 269071) and just 1 kilometer from the Hoscyn erosion grids, records an average annual rainfall of 1394 mms. The rainfall recorded during this experiment is graphed as Figure 16. 2.

The particle size distribution for the Waun Hoscyn spoil is indicated in Figure 16. 3. Once again, the spoil is D^{10}: 0. 2-0. 6 mms, and thus likely to have similar mechanical properties to that gathered at Waunafon and Milfraen.

However, there is one important respect in which the two instrumented slopes at Waun Hoscyn differ from those described at Waunafon. This is that these artificial slopes are not independent of the surrounding country-side. The Hoscyn spoil is banked up against the back wall of the Afon Lwyd's "230-meter" valley bench. Consequently, the spoil slopes are, functionally, just a small part of the much larger valley-side slope system. It is not clear how the opencast operation at Waun Hoscyn affected patterns of ground and surface water movement down the western slope of the Upper Afon Lwyd Valley. However, it is certain that the slopes of Waun Hoscyn receive a considerable through-flow,

P. 16. 2. --Sheep-cropped, Seeded Turf on Site of Waun Hoscyn V. Erosion Grid

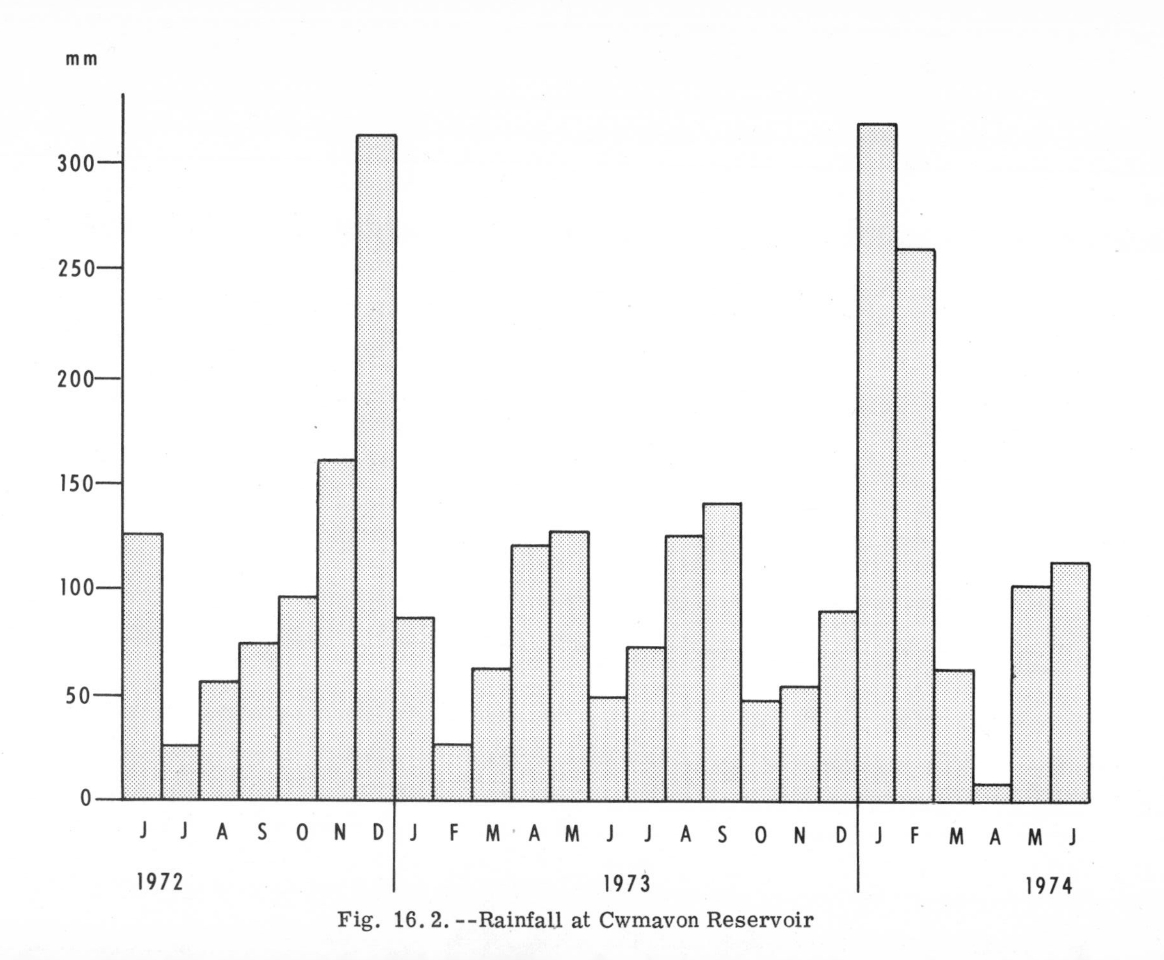

Fig. 16.2.--Rainfall at Cwmavon Reservoir

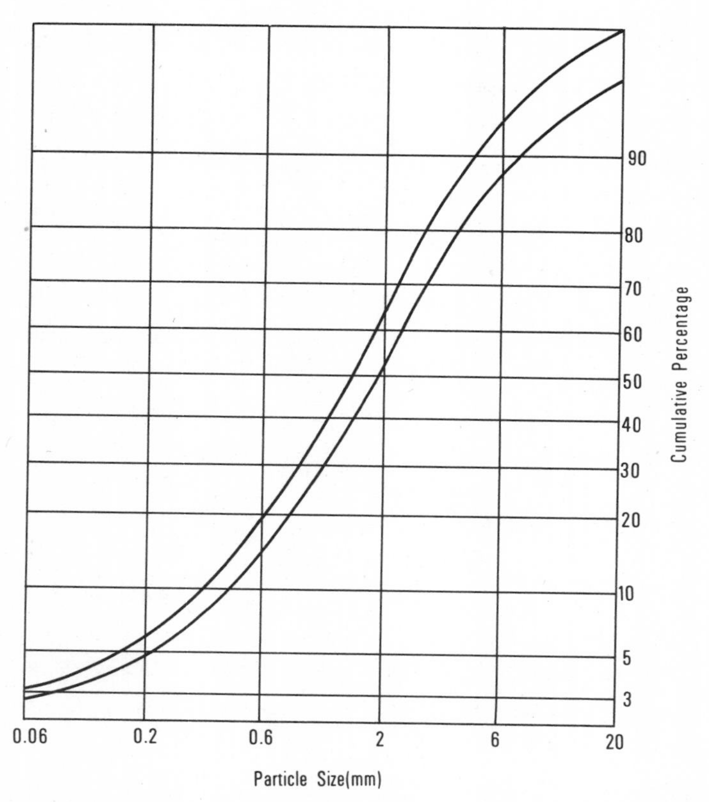

Fig. 16.3.--Particle Size Distribution--Range of 3 samples from Waun Hoscyn

as well as surface flow contribution from the undisturbed hill-sides upslope.

The experiments at Waun Hoscyn were established in July, 1972. However, much of Hoscyn "U." grid was destroyed by vandals during August and so the experiments were not effectively started until September, 1972, when the lost equipment was replaced. The analyses in this discussion are therefore concerned with the two-year period from September 1972 to September 1974.

The instrumentation employed at Waun Hoscyn is essentially the same as that employed in the studies at Waunafon and Milfraen. Two lengths of erosion pin are employed: 610 mms and 153 mms. These are established in meter wide rows at selected distances upslope. Each meter wide row is bounded by two 610 mm erosion pins and has three 153 mm erosion pins spaced out at 250 mm intervals between. Each erosion grid row at Waun Hoscyn has been carefully located so that it extends laterally into a neighboring gully channel. Typically, in each grid row, three erosion pins are situated on an inter-gully slope site, one or two on the gully banks, and one on the gully channel floor. Additional erosion pins have had to be added to particular erosion grid rows to accomplish all these requirements within the constraints of the fixed spacings allowed. The grid rows themselves are located at intervals on the slope which are multiples of a six meter unit, and are grouped more closely on those sections of the slope where the slope angle is changing most rapidly.

A general picture of results gathered from the inter-gully slopes of Waun Hoscyn is presented as Figure 16.4. The graphs chart the cumulative increase of the quarterly ground retreat estimates calculated from each slope profile. The implications of this graph are quite clear. The slope which is sparsely vegetated is retreating at a rate which is far greater than that recorded on the parallel vegetated slope. Thus, while the vegetated profile of Hoscyn V. is suffering a mean retreat of <u>circa</u> 3 mms p.a., the unvegetated slope of Hoscyn U. is suffering a mean retreat of nearly 10 mms p.a.

The detail disguised within the weighted average figures employed in the construction of Figure 16.4 is described in Figure 16.5 and Figure 16.6. These graphs record the average retreat recorded by the interfluve erosion pins in each grid row of each instrumented profile. It can be seen that the vastly greater average retreat recorded for Hoscyn U. is mainly a function of the high retreat scores from this grid's upper slopes. The maximum retreat records from Hoscyn U. were recorded by the grid rows at 60 meters (32.6 mm), 36 meters (28.5 mm) and 48 meters (23.4 mm). The maximum retreat scores from Hoscyn V. are more scattered and, of course, much smaller. The maxi-

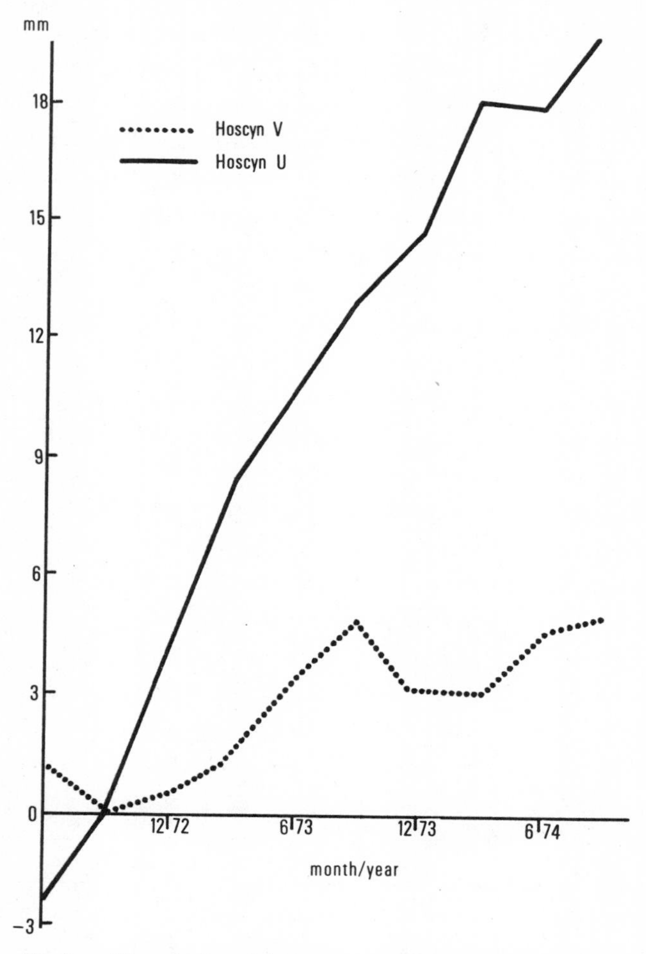

Fig. 16.4.--Mean Retreat of Waun Hoscyn Erosion Grids, 1972-1974

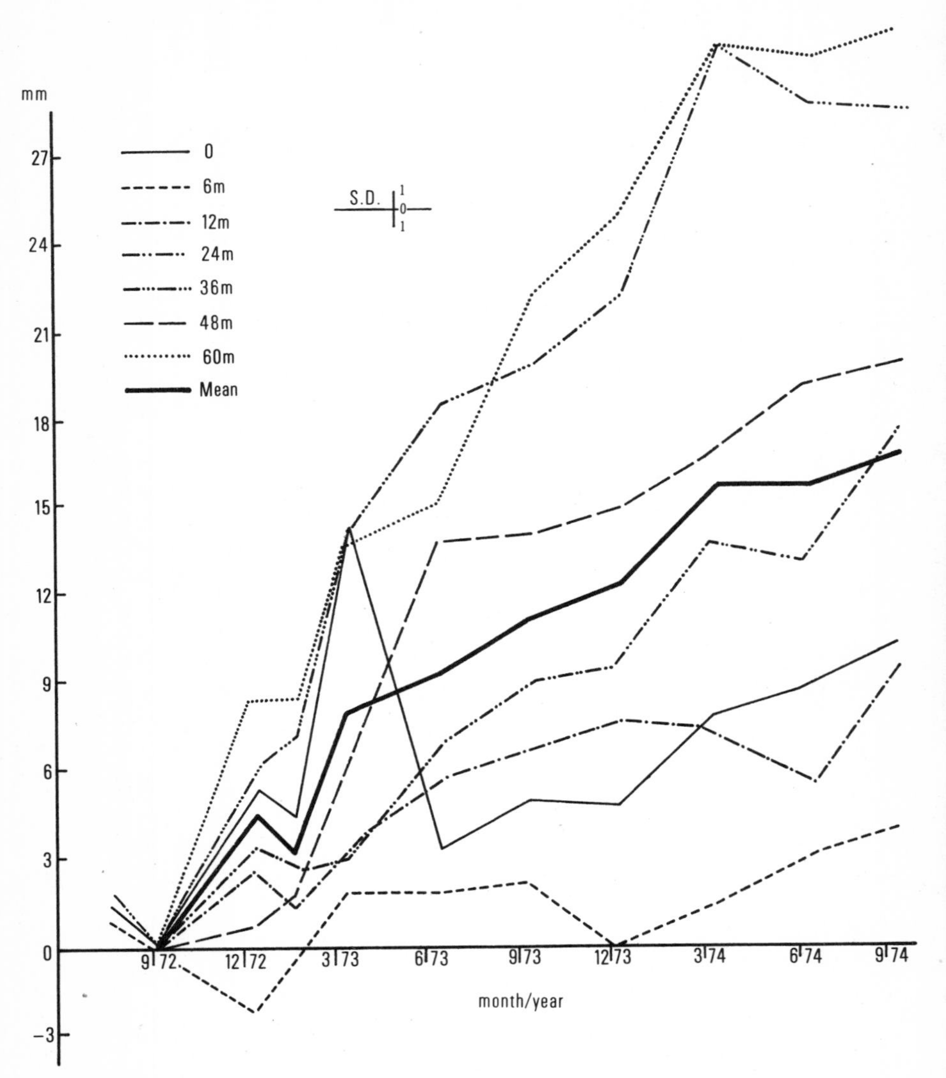

Fig. 16.5.--Waun Hoscyn U.: Ground retreat per grid row

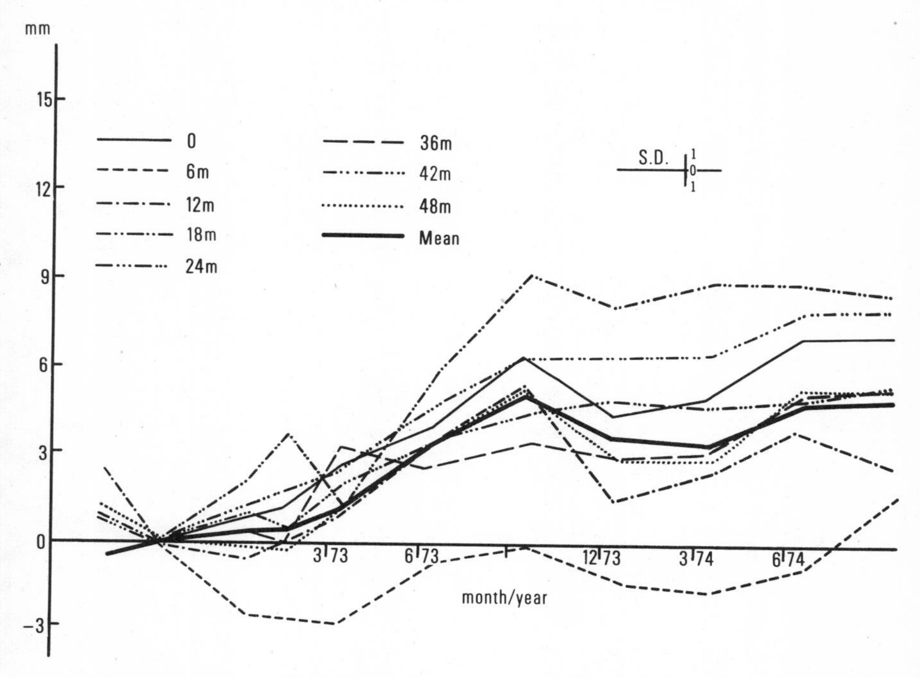

Fig. 16.6.--Waun Hoscyn V.: Ground retreat per grid row

mum retreat record here is the 8.5 mms recorded from the 18-meter grid row.

Both of the Waun Hoscyn slopes are affected by the existence of a gully near the slope foot. These gullies are not immediately adjacent to the actual slope foot but may well, still, be a factor in the reduction of slope foot deposition. Their effect on the pattern of recorded retreat is easily demonstrated by a consideration of the two slope foot traces: 0 meters and 6 meters on both Figure 16.5 and Figure 16.6. In both diagrams the 6-meter trace is the graph which records least net ground retreat. On both slopes the 6-meter grid row is that which is located nearest to the actual slope foot. By contrast the 0-meter trace features prominently in both diagrams. On Hoscyn V. (Figure 16.6), it is the grid row which suffers the third highest retreat (7.2 mms). Similarly, had it not been for one single massive phase of deposition during the spring of 1973, the 0-meter trace would have shown one of the higher retreat scores on Hoscyn U. (Figure 16.5). On both slopes the 0-meter grid row is established some distance from the slope foot upon a "pediment-like" feature, and near the basal gully channel.

It is easier to assess the implications of these results for slope development when the data is graphed in another way. In Figure 16.7 ground retreat recorded in each three month period is graphed against upslope distance. Since the two slopes under examination have a very similar morphology, the results from both are here represented on the same graph.

It can be seen that there is little correlation between the behavior of the two slopes during the experimental period. It is again emphasized that the increased retreat of Hoscyn U. is largely a measure of the activity of the upper slopes. The rates of retreat recorded on the lower slopes of the two grids (i.e. below 24 meters) are relatively similar. This situation suggests that there may be a fundamental difference in the patterns of slope development on the two slopes.

Table 16.1 displays the annual cumulation of quarterly ground retreat estimates expressed independently for the upper and lower sections of both slopes and both years of the experiment. The differences between the upper and lower slopes are expressed as the ratio of retreat on the upper slopes against retreat on the lower slopes. Once again, these ratios may be interpreted as follows: a ratio of less than 1:1 implies that the slope is becoming steeper and a ratio of more than 1:1 implies that the slope is becoming less steep.

Table 16.1 demonstrates that while the upper and lower sections of the vegetated slope are retreating at much the same rates, the upper slopes of

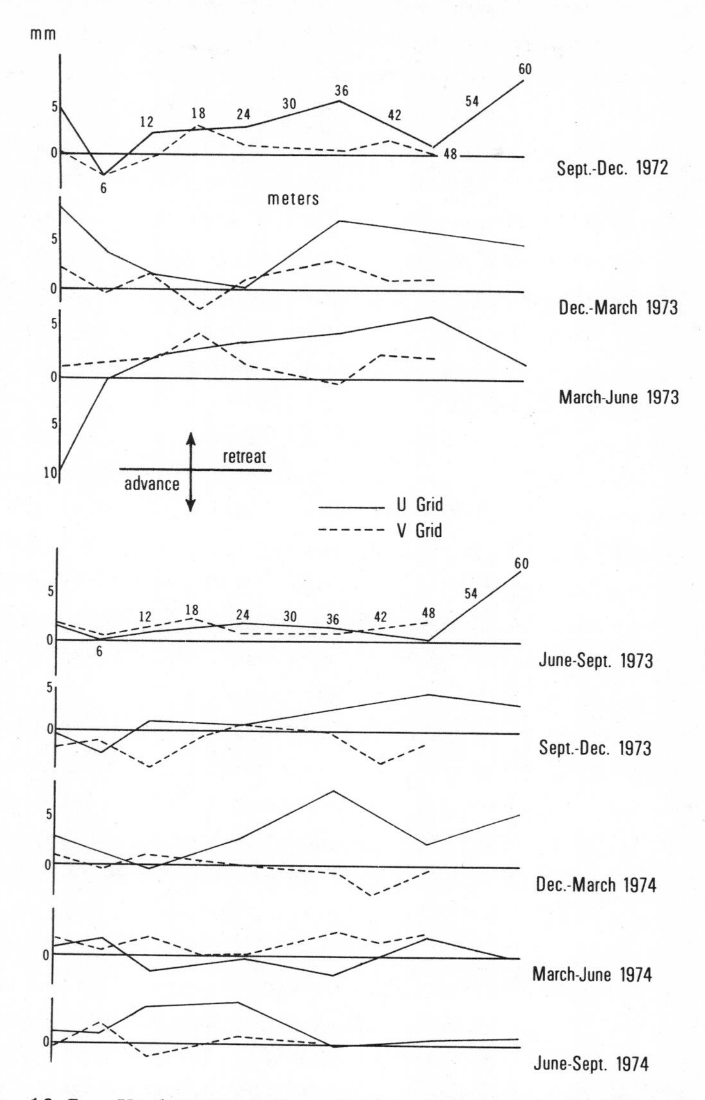

Fig. 16.7. --Upslope Variation of Ground Retreat in Each Quarter Year on Both Waun Hoscyn Erosion Grids.

TABLE 16.1

THE RETREAT OF UPPER AND LOWER SLOPES--
WAUN HOSCYN

Dates	Slopes	Retreat in Mm	
		Hoscyn U.	Hoscyn V.
1972-1973	Upper	18.99	4.99
	Lower	6.63	5.19
	Ratio U/L	2.86	1.04
1973-1974	Upper	8.71	-0.40
	Lower	4.62	-0.38
	Ratio U/L	1.89	0.95
1972-1974	Upper	27.70	4.59
	Lower	11.25	4.81
	Ratio U/L	2.46	0.95

Hoscyn U., the unvegetated slope, are retreating approximately 2.5 times as fast as the corresponding lower slopes. If this situation in fact constitutes a general trend, as a consideration of Figure 16.5 and Figure 16.7 might suggest is possible, then it can be concluded that the slope evolution regime of the sparsely vegetated Hoscyn U. slope is characterized by slope decline. In the same way the approximately 1:1 ratio of retreat on the upper and lower slopes of the vegetated slope Hoscyn V. implies that slope evolution here is dominated by parallel retreat.

Some estimation of the strength of these differences is revealed by a comparison of the retreat records from the upper slopes and the lower slopes of both erosion grids. Thus, while retreat on the upper slopes of Hoscyn U. exceeds that recorded on the comparable upper slopes of Hoscyn V. by a factor of 6.04. Retreat on the lower slopes of Hoscyn U. is only 2.34 times that recorded on comparable sections of Hoscyn V. grid.

In sum, it can be seen that the presence of the dense seeded turf on Hoscyn V. grid plays a very important role in the restriction of erosion and ground retreat. Rates of slope evolution on the sparsely vegetated Hoscyn U. grid are higher than those recorded on Hoscyn V. by a factor of 3.85. The pattern of slope evolution is also substantially different. The densely vegetated Hoscyn V. is evolving mainly by parallel retreat. The sparsely vegetated slope of Hoscyn U. is evolving by slope decline.

Although the slopes themselves have a comparable morphology, the gullies which are developing upon them do not. The small secondary gullies of the densely vegetated sections of Waun Hoscyn are characteristically very narrow and fairly deep. They possess an angular box-like cross-section. Their average depth is of the order of 0.5 meters and their width is about the same. Secondary gullies developing on the sparsely vegetated slopes have a rather different character. Here there is no turf to help preserve the steep gully sides and as a consequence these gullies tend to be much wider and much shallower. The gullies of the vegetated slopes seem to receive much of their water and sediment from pipes developing under the turf mat. However, on the less vegetated slopes a sheet-washed surface which surrounds the scattered grass tufts grades, almost imperceptibly in places, into the wide gully wash.

A crude comparison of the results from the gully channel, gully bank, and gully interfluve erosion pins is presented on Figure 16.5 and Figure 16.6. Here, three lines representing the mean retreat per gully interfluve, gully channel, and gully bank erosion pin have been super-imposed on the graphs of ground retreat per (interfluve) grid row.

The results are a little surprising, for by this record, there is no active gully incision on either slope. The graphs from Hoscyn U. (Figure 16.5) show that the mean retreat recorded by interfluve erosion pins was twice that recorded by erosion pins in the channel floor, but that interfluve retreat was exceeded by bank retreat. This implies that the gully under examination is in the process of disappearing into the sheet-washed surface which surrounds it. The results from Hoscyn V. (Figure 16.6) have a similar message. Here, the records from the gully channel are nearly neutral. However, gully bank retreat proceeded at a respectable rate though the rate of interfluve retreat was slightly greater. The difference, however, was probably not significant in terms of channel widening.

Figure 16.8 is a record of the total retreat recorded by the erosion pins in each grid row, in gully channel, gully bank, and gully interfluve situation. Results from the two instrumented slope profiles are graphed separately. It can be seen that the retreat of the gully interfluves was greater than either the retreat of the gully banks or the gully channel over most of the slope profile on both slopes. Interfluve retreat was exceeded by gully bank retreat at either extreme of the two gully channels. Gully channel retreat exceeded interfluve retreat at either extreme of the Hoscyn U. channel, but on Hoscyn V. it was everywhere less than interfluve retreat. Bank retreat was only exceeded by channel retreat over a small section of the lower gully thalweg.

These results suggest that the gully channels of both slopes are gradually

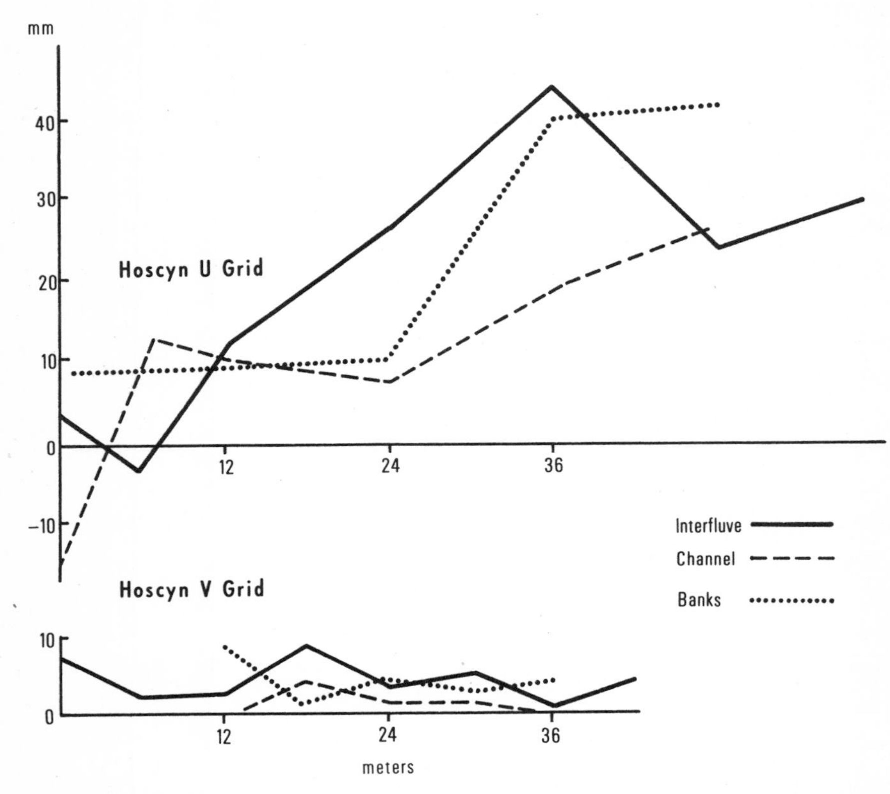

Fig. 16.8.--Gully Incision at Waun Hoscyn, September 1972-September 1974

disappearing. Active gully incision was only a feature of the upper and lower extremes of the Hoscyn U. gully, though active bank erosion was a feature of the extremes of both gully channels. (The implication contained by Figure 16. 5, that bank retreat generally exceeded interfluve retreat, was a consequence of an attempt to use a mean figure to describe clustered data. Grid rows are more densely spaced on the lower slopes of both grids.)

Figure 16. 9 reduces the generalization of Figure 16. 8 to its separate components. These graphs depict, separately, the retreat recorded at gully bank, channel and interfluve sites during each season of the year on each slope. Figure 16. 4 established that Hoscyn U. suffers its greatest interfluve retreat in the winter months and Hoscyn V. its greatest interfluve retreat in spring. Figure 16. 9 demonstrates that this pattern is not echoed by retreat in the gully sites and that there are notable secular and spatial variations in retreat occurring through the experiment.

It can be seen that gully bank retreat is mainly a feature of the early part of the experiment, especially on Hoscyn U. Bank retreat seems to have reached a peak of activity on the upper slopes during spring 1973. Later bank retreat seems to have been on a much smaller scale and was restricted to the lower slopes and slope foot.

Gully incision on Hoscyn V. (i. e., when gully channel retreat exceeds interfluve retreat) was mainly a feature of the summer and autumn months. However, on Hoscyn U., this seasonality was not apparent. Incision here appears to escalate slowly throughout the experiment. It first affects the slope foot in summer 1973, then extending during 1973 attains a maximum in spring 1974 when incision affects the lower 36 meters of the slope profile. Later records show progressively less and less incision on the lower slopes, but the commencement of incision on the upper slopes. The slopes around the 48-meter grid row are the first to be affected in spring 1974, but incision extends to affect all of the slopes above 30 meters in the following summer.

Both slopes are dominated by the condition of ground retreat. However, ground advance records are not uncommon especially on the less active Hoscyn V. Ground advance was a feature of the lower interfluve slopes during both autumns and a feature of the upper interfluve during both autumn 1973 and winter 1974. Channel floor advances were widespread across the upper profile throughout the period March 1973-March 1974. However, major ground advance records from the gully banks are mainly restricted to the first six months of the experiment. Records of ground advance on Hoscyn U. are mainly

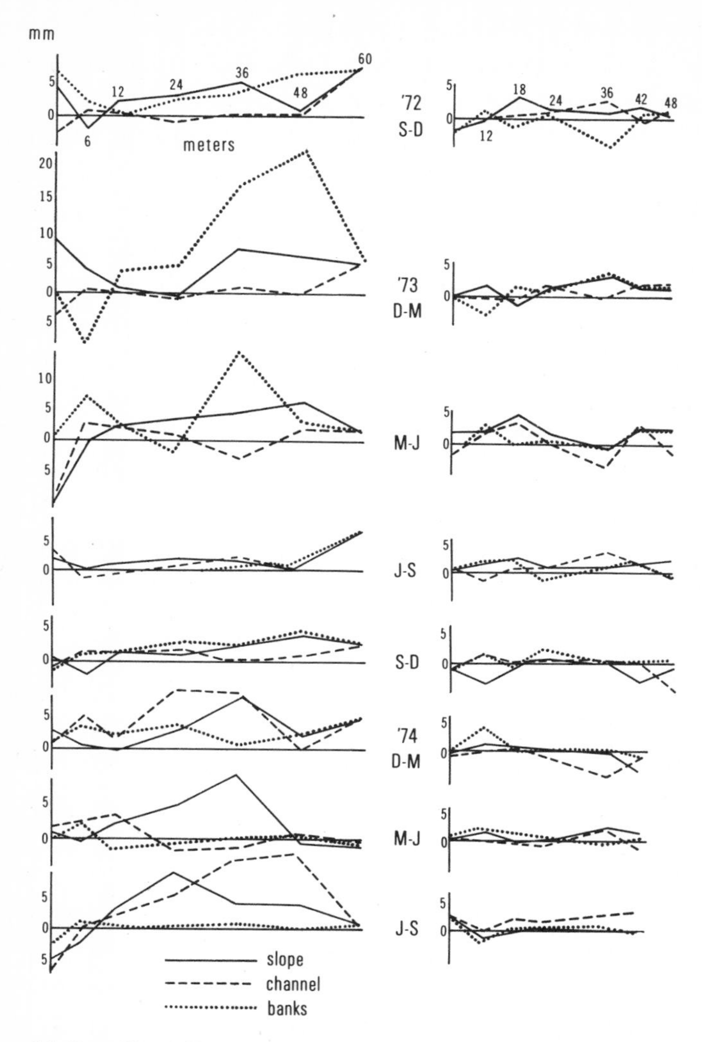

Fig. 16.9.--Waun Hoscyn: Comparison of quarterly ground retreat records for slope profile, gully banks and channel sites.

restricted to the slope foot, and are more common from the erosion pins sited in the gully. Advance records from other parts of the profile are uncommon, though parts of the gully banks and channel register ground advance during both springs of the experiment.

Figure 16.10 is an attempt to interpret the sum of all these results in terms of actual slope evolution. Here, an exaggerated representation of the average annual ground retreat recorded on the inter-gully slopes and in the gully channels of the instrumented profiles of Hoscyn U. and Hoscyn V. erosion grids, has been constructed on a straight line graph of the actual slope profile. This straight line has been bent to conform with the calculated configuration of the mean, flow line slope profile of the erosion grid.

The dimensions of the actual retreat records from Waun Hoscyn are, of course, not of the same order of magnitude as the gross slope morphology. The visual impact of Figure 16.10 is achieved by multiplying average annual retreat by a factor of 100. The diagram may thus be interpreted as a prediction of slope evolution during the next 100 years, provided that one also accepts the assumption that the results measured during 1972-1974 are representative of that period.

Figure 16.10 also emphasizes the differences in the rates of retreat experienced on the two slopes. It can be seen that Hoscyn V. is undergoing very slight retreat. In fact, most of the slope is retreating parallel to itself and the main projected morphometric change is a slight extension of the slope foot "pediment" and a slight steepening of the main slope. The alterations suggested for the slope which is not protected by a thick seeded turf are far more dramatic. Hoscyn U. will, on present form, suffer major slope decline and a marked extension of its slope foot "pediment."

In the diagram, the dashed line represents inter-gully slope retreat and the dotted line represents gully incision. It can be seen that across most of both slopes the dashed line lies well below the dotted line. The distance between the two lines represents the decline of the visible depth of the gully channel. Where this exceeds the actual depth it amounts to a prediction that the gully will have disappeared. Figure 16.11 permits the suggestion that the Hoscyn U. gully channel will have largely disappeared by the year 2072.

It must be emphasized, however, that these conclusions apply only to the small secondary gullies on each slope. There is plenty of evidence that the larger master gullies on both slopes are very active. Photograph P.16.3 depicts a master gully developing near Hoscyn V. grid. This gully has achieved

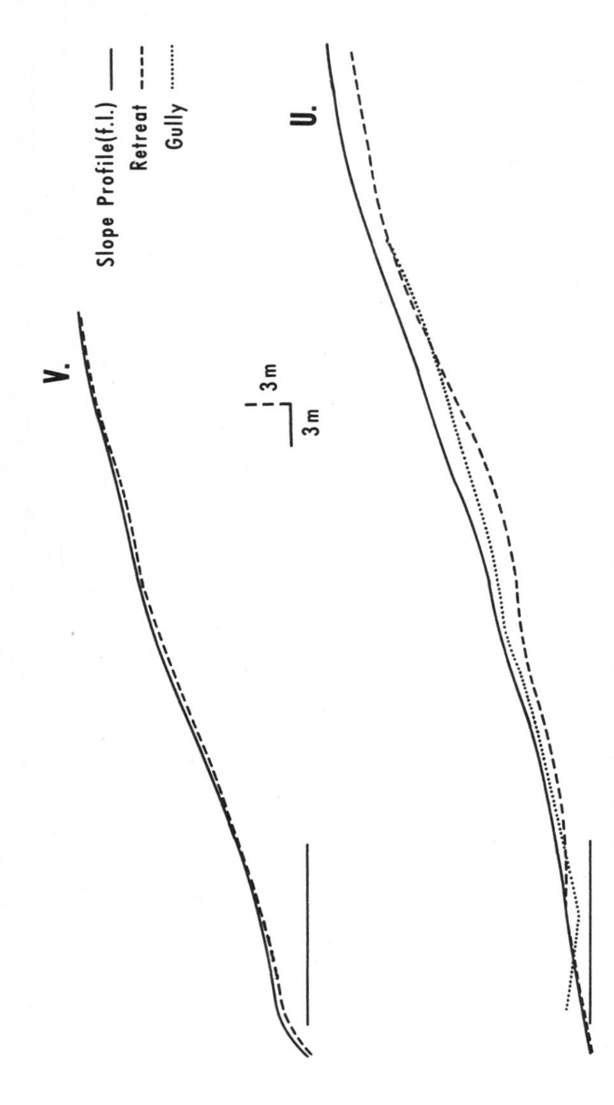

Fig. 16.10.--Annual Slope and Gully Retreat x 100 at Waun Hoscyn

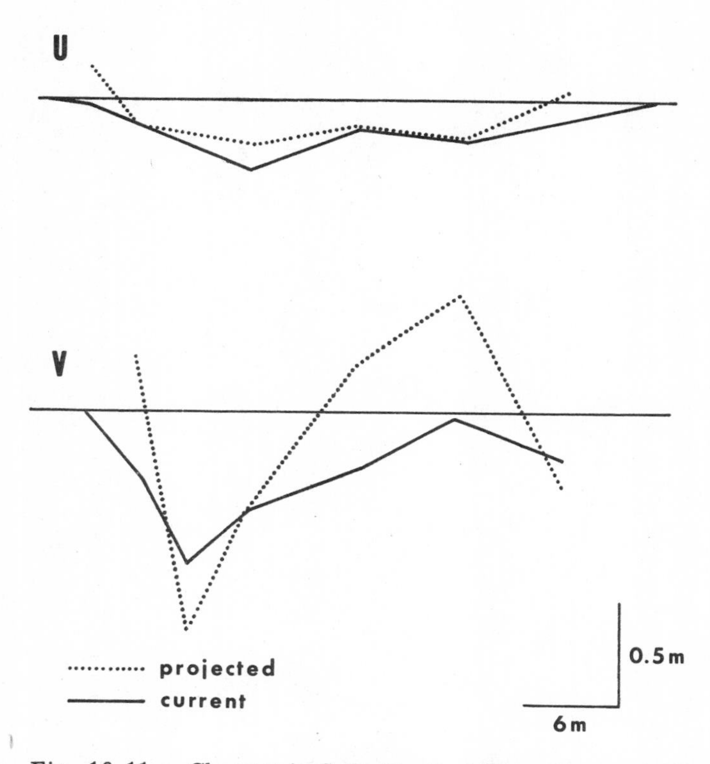

Fig. 16.11.--Changes in Gully Depth at Waun Hoscyn (x25)

P. 16. 3. --Master Gully near Waun Hoscyn V. Erosion Grid

its mean depth of 3 meters in just twenty years. Photograph P. 16. 4. witnesses that the situation on the vegetated slopes near Hoscyn V. grid are little better. This gully has a depth of circa 2 meters. The photograph depicts one of four surviving bank erosion pins established flush with the gully's bank in July 1972. The washer was reset flush to the bank in July 1974 and the photograph was taken in the following September. The retreat recorded by these two pins during the two years July 1972-July 1974 were, from the south facing gully bank, 117 mms and 91. 5 mms and, from the north facing gully bank, 220 mms and 89. 5 mms. These are the largest retreat scores measured from any situation on any of the artificial landforms considered by this study. They suggest most strongly that the control of gully erosion must be a major feature of any land reclamation project.

It is possible to give an approximate date to the termination of the existing conditions of gully erosion at Hoscyn V. grid. At present, this slope is protected from any direct gully control by its turf-covered foot slope. This extends some 6 meters from the present gully bank. As the gully banks retreat they must reduce this gap and increase the possibility of a basal rejuvenation of this quiescent secondary gully. If one assumes a mean bank retreat rate of 100 mms p. a., then in 60 years the gully bank will intercept the gully channel and give the channel an effective 2 meters of additional relief. This of course assumes that this process is not accelerated as the proximity of the master gully increases and as the slope foot accumulations of Hoscyn V.'s gully slowly disappear. The behavior of the 0-meter trace on Figure 16. 7 however, suggests that some acceleration of ground retreat must inevitably occur.

Normally, in the absence of a basal gully control, one should expect the slope foot to record ground advance because of the accumulation of eroded materials. At Waun Hoscyn, the 6-meter grid row is that which tends to record least actual retreat. This is located precisely at the slope foot. The 0-meter grid row is located some way beyond this slope foot near to the basal gully channel. Accordingly, the 0-meter trace on the graphs of Figures 16. 4-16. 10, show a greater activity than those of the 6-meter, slope foot, trace. This increased activity must surely, in some measure, be a function of the proximity of the basal gully's bank. The interception of basal gully bank and secondary gully channel will cause an immediate rejuvenation of the secondary gully channel.

In sum, it can be seen that the presence of a dense turf causes a major reduction in effective ground retreat. Ground retreat on sparsely vegetated sec-

P. 16. 4. --Bank Erosion near Waun Hoscyn V. Erosion Grid

tions of the Waun Hoscyn infilled opencast site was nearly 10 mms p.a., but that on the densely vegetated sections was only 3 mms p.a. Slope evolution on the densely vegetated slopes was dominated by a tendency towards parallel retreat modified by slight extensions of the foot slope and slight mid-slope steepening. Slope evolution on the sparsely vegetated sections of Waun Hoscyn was dominated by slope decline and a major extension of the foot slope. Gully incision was only active at slope foot and head-ward sites on the sparsely vegetated slope. Bank retreat was most significant at either extreme of the gully channels. There was a net decline in the depth of the secondary gullies studied by this experiment. Measurements of bank retreat on the large continuous gullies of this site suggest that this situation will be changed in the foreseeable future, when the bank retreat of these master gullies may effect a rejuvenation of the presently quiescent discontinuous secondary gully channels.

CHAPTER 17

GROUND RETREAT AND SLOPE DEVELOPMENT ON SURFACE MINING SPOIL BANKS NEAR BLAENAVON

The erosion of opencast mine-fill spoil banks is a subject of increasing and world-wide concern. Opencast operations in Britain produce a relatively small proportion of the nation's coal, but 39% of the coal extracted in the United States (Rushton - 1973) and 28% of the coal extracted in the U.S.S.R. (Brealey and Atkinson - 1968) is produced from surface mines. A recent survey of the 3.2 million acres in the United States which have been affected by surface mining operations revealed that about half the sites contributed significantly to sediment pollution in the surrounding environment. Further, that while 60% of the sites did not suffer from any serious gully erosion, 10% possessed gully channels which were more than 0.3 meters deep (U.S. Dept. Interior - 1967).

This study has been concerned with gullying and slope development on two upland "problem" opencast sites. The features these two sites share are severe gullying, erosion, and devegetation problems, and proximity to the town of Blaenavon. They are separated by two dissimilarities. The first is that while one, Waunafon, is essentially independent of the surrounding landscape because of its location on the crest segment of the Afon Lwyd's valley slope, the second site, Waun Hoscyn, which is located near the backwall of a well developed valley bench, is, functionally, just a small part of the larger valley-side slope system. The second difference is that while Waunafon was never subjected to a formal restoration, the site at Waun Hoscyn is an example of a site whose initial restoration has failed.

Instrumented slopes have been established on both sites and records of ground retreat have been collected for two years. Five profiles have been monitored at Waunafon. All are located in the artificial valley at Cefn Garn-yr-erw and all are subject to a basal gully control. Two of the profiles have been established on vegetated slopes of opposed aspect; one faces east, the other faces west. Two further profiles are located on unvegetated east facing slopes and one of these is established in close association with an instrumented secondary

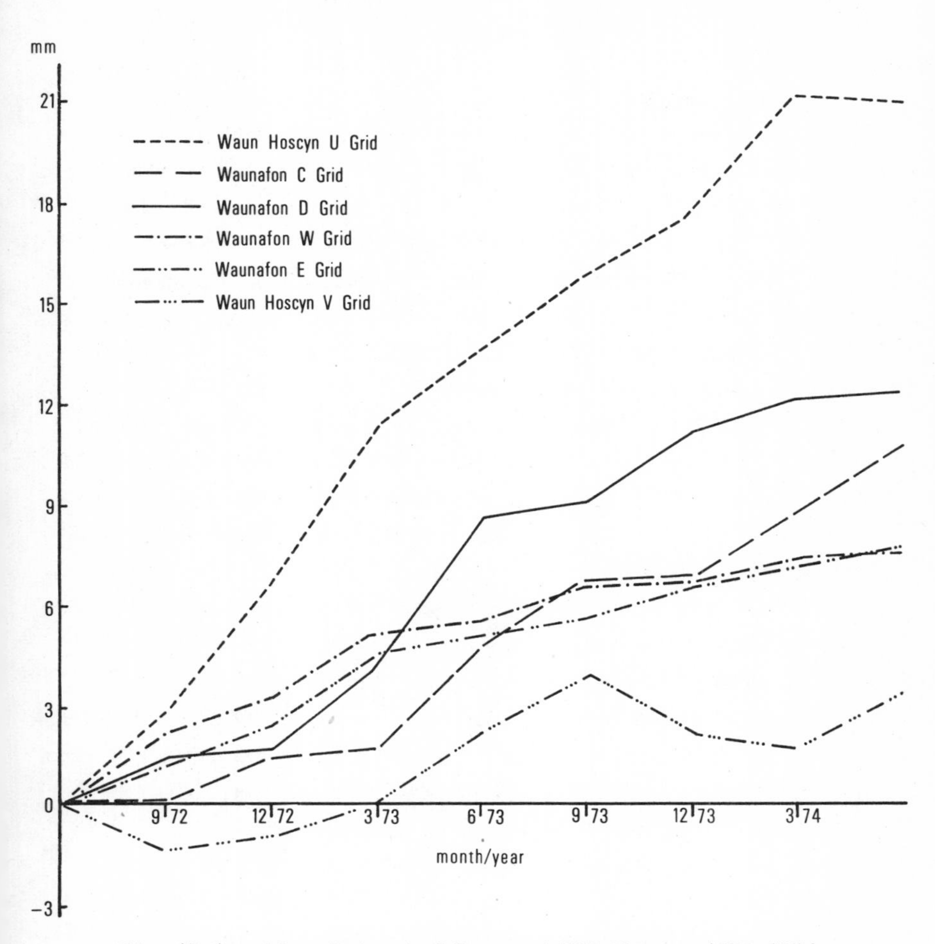

Fig. 17.1.--Mean Retreat of Opencast Fill Slopes: 1972-1974

gully profile. Two erosion grids have been established at Waun Hoscyn and each includes interfluve, gully bank, and gully channel sites. One erosion grid is located on a slope covered by a dense seeded turf, the other is very sparsely vegetated. The two grids are located within 300 meters of each other on the same east facing slope.

The results from all the instrumented interfluve slopes is displayed as Figure 17.1. It is shown that the retreat of those slopes which lack vegetation is far greater than those which are vegetated. The two extreme results are provided by the data from Waun Hoscyn. The seeded turf of this vegetated slope (Hoscyn V.) is a far more dense vegetative mat than that found on the naturally revegetated slopes at Waunafon (Waunafon E., Waunafon W.). Consequently, the rate of ground retreat recorded at Hoscyn V.--2.34 mms p.a.--is far lower than that recorded on the two vegetated grids at Waunafon: Waunafon E., 3.77 mms p.a.; Waunafon W., 3.75 mms p.a. The relatively large rate of retreat experienced on the sparsely vegetated slope at Waun Hoscyn, Hoscyn U.: 9.52 mms p.a., compared with the results from the two unvegetated slopes at Waunafon--Waunafon C., 4.78 mms p.a.; Waunafon D., 6.12 mms p.a.

There is a sharp contrast in the patterns of slope evolution recorded on the vegetated and unvegetated slopes. The slopes which are clothed by vegetation all display a tendency towards main slope evolution by parallel retreat, and little apparent profile modification except by the extension of the lower slope element. The pattern of slope evolution on the unvegetated slopes is more complicated and is in each case suggestive of major profile modification. The simplest pattern of alteration is projected from the data gathered at Hoscyn U. Here there is a progressive upslope increase in ground retreat records which implies that this slope will evolve by main slope decline. The main-slope records from the environment-independent unvegetated slopes of Waunafon show that parallel retreat main slope is supplemented by a more rapid retreat of the upper convex element.

CHAPTER 18

CONCLUSION

Man is the most active exogenic geomorphological agency. Artificial landforms dominate a large proportion of the world's dryland surfaces and the rate of anthropogenic landform generation accelerates each year. The study of man-made landforms is, however, a neglected area in geomorphology.

The particular concern of this study has been an examination of landscape development on directly created constructional landforms derived from colliery spoil materials and produced in the course of deep and surface mining operations. The study is set in the old industrial district around Blaenavon, Gwent, at the head of the "Eastern Valley of Monmouthshire" and on the north-eastern outcrop of the South Wales Coalfield. The Blaenavon district is an area which has suffered continuous and intensive industrial exploitation for two centuries and which continues to carry most of the scars. The area contains a wide range and large number of artificial landforms derived from local materials by different technologies at different times.

The study of slopes is an important and central theme within geomorphology. Most landforms may be considered as combinations of different slopes. Unfortunately, the study of slopes has yet to progress far beyond a few, usually unsubstantiatable, theoretical speculations and a limited understanding of the operations of some of the more important slope processes. However, a consideration of some recent empirical investigations into the relationship between slope form and soil loss, when viewed in the light of results from the investigation of individual slope processes, allows certain limited conjecture concerning the nature of slope evolution tendencies.

The artificial landforms at Blaenavon are of known age, construction, and history. They are of comparable size, mainly between 10 and 20 meters local relief, and usually, simple form. They are being affected by slope processes and their controls in a variety of ways. These slopes, therefore, are employed as a frame-work for certain limited investigations into the nature of natural slope evolution. Since the variations in erosional environment compassed by these experiments are also important to considerations of the evolution of the artificial landforms themselves, the dual aims of this study are not mutually exclusive.

The artificial landforms considered by this study are of two fundamental types: either they are created from the spoil produced during deep mining or they are created from opencast mine fill. The deep mine spoil banks tend to be discrete landforms which cover a small area. The surface mine spoil banks tend to be less obtrusive but occupy larger areas. Deep mine spoil tends to consist mainly of coal-shales and may have a high carbon content. Opencast fill tends to be far more heterogeneous and reflect more closely the composition of the strata overlying the coal-bearing seams.

The main experiments contained by this study involve the direct measurement of morphological change. Erosion pins were established in fixed format rows down flow-line slope profiles at selected sites. Fixed format rows contain two types of erosion pin: 610 mm and 153 mm. Each row is one meter wide and bounded by two 610 mm rods. Three 153 mm rods are spaced out between. Erosion pins are allowed an initial exposure of circa 15 mms and fitted with a washer to facilitate data collection. Rows are established such that any vertical or lateral displacement of individual pins will be apparent. Rows are spaced at intervals down the profile which are multiples of a 1.5-meter unit. They are clustered in areas where the change of slope is most rapid.

The two varieties of deep mine spoil mound examined were loose-tipped spoil cones and the smaller tips created by tipping and compaction. Ground advance at the foot of the loose-tipped cones proceeded at a rate of 5-8 mms p.a. Mass movements, especially creep and small scale sludgings, were important processes here. The surfaces of the compacted tips were far more stable and mass movements were restricted to a thin surface layer or to local terracette generation. Most of these tips possessed a vegetative cover though this tended to be thinner on the steeper slopes. The presence of a vegetative cover restricted mean ground retreat to 3 mms p.a. Devegetation usually commenced with the development of a bare scar at the upper convexity and this development was not entirely independent of the grazing, trampling, and burrowing activities of sheep (Photograph P.18.1). The tendency for increased erosion at the upper convexity was also encouraged by the initial dump morphology which provided all these slopes with an abrupt break of slope separating the main slope and crest segments. This discontinuity was universally highlighted as a zone of increased erosion and morphometric alteration both by the direct measurement studies and by supportive morphometric analyses of other compacted spoil tips.

P. 18. 1. --Sheep as an Agency of Erosion--Waunafon

Devegetation, itself a consequence of increased surface instability, causes an acceleration of the rate of retreat to 5 mms p. a. If the devegetated slope exists in a location where the basal accumulation of detritus is prevented, overall retreat may be as much as 9 mms p. a.

The status of ground retreat was also determined on two opencast mine spoil banks, one of which was composed of slopes which were independent of inputs from the surrounding landscape, the other whose slopes were part of a much larger slope system and which received inputs of sediment and water. Both spoil banks were partially vegetated and both were finely dissected by gully channels. Most of the slopes on these sites possess some form of basal control, different from that recorded on parallel unvegetated slopes. Frequently, it seemed that the degree of vegetation of a slope had a greater effect on mean slope retreat rates than did the activity of the basal slope control.

Slopes which face south and west tend to retreat faster than those which face north and east. Even on opposed east and west facing slopes there was a tendency for differential slope steepening which might be attributable to aspect.

The effect of the basal control on slope development tended to be greatest nearest the actual channel. The absence of any basal removal of rock waste resulted in accumulation on the lower slopes and the extension of the slope into the surrounding landscape. The importance of the basal control was greatest on slopes which were directly undercut. Variations in the conditions of removal at the foot of such slopes were more or less directly communicated upslope. However, it was discovered that there was a tendency for many basally controlled slopes to develop and maintain lower slope elements which separate the main slope segment from the parts of the slope which are directly influenced by the basal control. This lower concave element appears to buffer the main slope from the effects of short-term variations in the conditions of erosion or deposition at the slope foot. The basal control usually appeared to be especially active at the foot of the unvegetated slopes proceeded at a rate of between 5 and 6 mms p. a., except on the unvegetated slope which was a part of the larger slope system. Here, the rate of retreat was much greater at 9. 5 mms p. a. However, a section of the same slope which had been reclaimed and which supported a dense seeded turf retreated at a much slower rate--only 2. 3 mms p. a. The retreat of the environment-independent slopes which were covered by the far thinner natural vegetation proceeded at 3. 8 mms p. a.

Study was also made of the evolution of four secondary gully profiles developing on the artificial landforms of Blaenavon. These gullies seemed to belong to two categories: those which were integrated into the main drainage network and those which were essentially independent. The gullies which were connected to the master-gully network by a well developed channel were highly active in erosion and transportation. Those which were not so connected were found to be mainly inactive.

Recent developments in slope research suggest that, if the possible complications afforded by a consideration of the effects of the unit-weight variable are ignored, then soil loss by rain water erosion is greater on longer slopes of convex profile. Concave slopes suffer least from erosion. Water erosion, however, is a complicated process which is affected as much by the physical properties of the soil as by the external agencies which effect erosion. However, water erosion is not the only process system involved since creep and deflation may be of importance. The operations of the slope process systems were inhibited by the presence of vegetation, but encouraged by disturbances due to animals. Slopes which faced south and west were shown to be more susceptible to retreat than those which faced north and east, at least in Blaenavon's present climatic context. Further, it was demonstrated that the activity of the slope's basal control was critical to its mode of development.

Slope evolution on the colliery spoil mounds was dominated by the retreat of the upper convex slope element and its extension into the crest segment and the main slope segment. The main slope segment suffered parallel retreat except near the slope foot where this process was liable to interference by occasional extensions of the slope foot accumulation regime. There was no evidence of basal accumulation on any of the slopes which possessed a basal fluvial control. However, there was a tendency for neutral slope records to be associated with the development of a concave lower slope element separating main slope from gully bank. Parallel retreat was the rule for all vegetated main slope and upper slope units and also for the main slopes of the environment-independent unvegetated profiles. The upper convex elements of these slopes displayed a tendency towards increased erosion. However, on the unvegetated slope which was subject to inputs from other slopes upslope, there was a steady increase in retreat upslope which suggests that this slope was evolving by decline.

CHAPTER 19

PROBLEMS OF STUDIES OF HILL SLOPE EVOLUTION BY DIRECT MEASUREMENT

Fundamentally, the direct measurement of geometric change is the least equivocal method of examining the progress of slope evolution. Measured changes are real and finite. The hazy areas of speculation which must inevitably cloud the reliability of studies which rely on simulation or morphometric comparisons are greatly reduced. Nevertheless, direct measurement does cause problems and these have to be examined.

Natural hill slopes evolve very slowly. The techniques usually employed for direct measurement tend to be either very crude, or cumbersome, or expensive and applicable to very small areas (cf: Campbell - 1970). The areas which may be instrumented in the course of a single research project tend to be limited enough in any case, so there has been a tendency for research workers involved in direct measurement to substitute quantity for sophistication in their instrumentation and to limit their efforts to the examination of slopes whose rate of development is sufficiently large to guarantee that the results gathered will be larger than the margins of error implicit in the particular technique employed. In practical terms this has meant that direct measurement studies have always concerned themselves with special slopes whose rapid evolution is, usually, a consequence of anthropogenic disturbance. The literature review in Chapter 7 illustrates this limitation. It can be seen that the majority of these studies are directed to the examination of North American "Badlands," agricultural slopes, or entirely artificial slopes. Now, it has been stated repeatedly throughout this study that such anthropogenic landscapes are, in fact, the normal landscapes of most of the world (Chapter 1). However, these anthropogenic slopes have a singular character and, strictly, should only be directly compared with certain extreme classes of natural landscape. An alternative approach has been suggested in this study where the artificial slopes examined are regarded as laboratory analogues of natural hill slopes on which the effect of certain naturally occurring slope process controls may be examined in idealized conditions (Chapter 2, Chapter 6).

However, not even the limitation of direct measurement studies to the world's most rapidly evolving slopes entirely overcomes the main problem of direct measurement studies, for the evolution of slopes is still a relatively slow and long term process.

The time factor continues to be a problem. Schumm's 1952 pioneer study of slope evolution on the infilled clay pit of Perth Amboy, New Jersey was based on just ten weeks data (Schumm - 1956a). Such a brief study would not be so credible today. The author's own research, which was designed to meet the constraints of a research grant from the Natural Environment Research Council of Britain, was budgeted for two years' data. This, too, proves to be a painfully short data record. Leopold and Emmett (1972) have reported results from 6 to 9 years of study at selected sites in the United States but even 6-9 years is not very long in terms of slope evolution. In fact, of current studies, only the British Association's work at the Experimental Earthwork at Overton Down has a really plausible time-scale, and that has a program of observations scheduled to extend over 100 years (Jewell - 1963; Crabtree - 1972).

Direct measurement tends to force the research worker into a slavishly Uniformitarianist stance. The interpretation of results, in terms of patterns of predicted slope evolution, inevitably involves a series of mammoth multiplications through either time and/or space. Possibly then, the most awkward criticism which one can direct to the author of a direct study is one which suggests that the years of measurement of his particular study are atypical, and that consequently, the results should not be multiplied or projected. The criticism is all the more cogent because there really is no such thing as a "typical" year.

Figure 19.1 is a graph depicting the ways in which the rainfall during 1972-1974 exceeded the range of the monthly rainfalls recorded previously. The data are taken from the Llanelly Hill rain-gauge which is located in the middle of the author's research area near Blaenavon (Gwent). There are considerable differences even between the rainfalls recorded in the years of the experiment.

Figure 19.2 depicts ground retreat data from two instrumented slopes on colliery spoil mounds near Blaenavon (Gwent) graphed along the slope profile. Figure 19.3 depicts ground retreat data gathered from a west facing slope developing on the opencast mine fill at Waunafon. The visual effect of these graphs has been achieved by multiplying the actual retreat values by a factor of one hundred. In each case, the results from the two years of the experiment

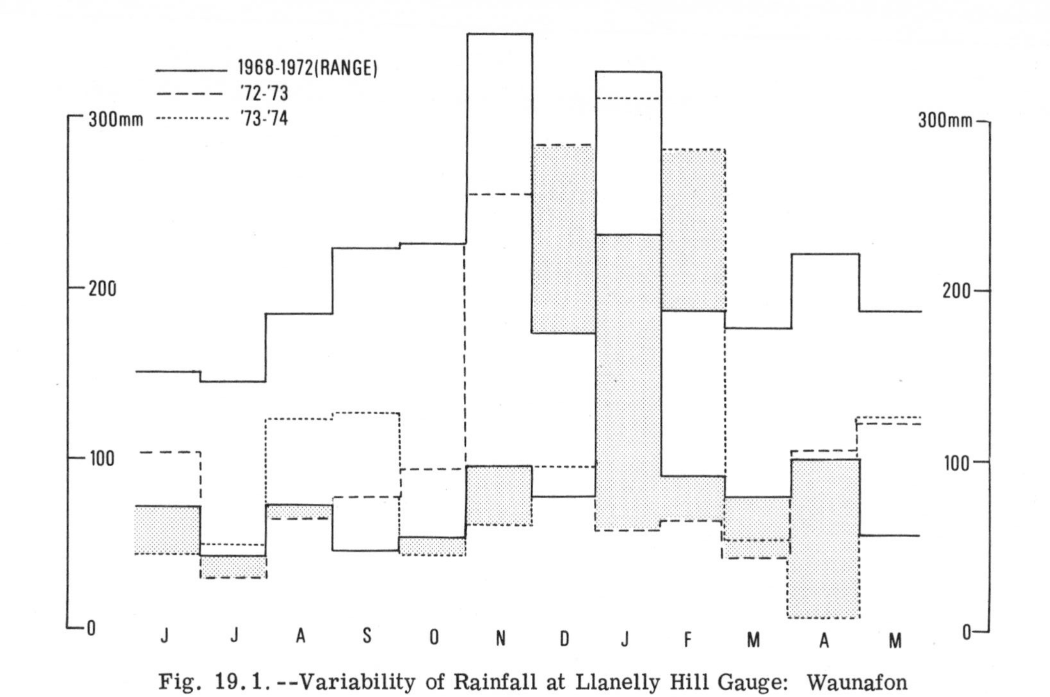

Fig. 19.1.--Variability of Rainfall at Llanelly Hill Gauge: Waunafon

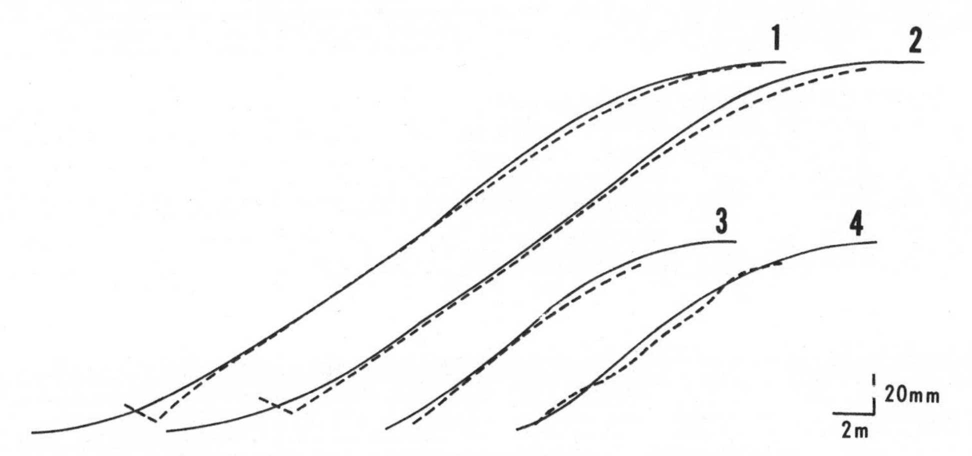

Fig. 19.2.--Milfraen Vegetated Slopes, Ground Retreat 1972-1974. On these graphs ground retreat x 100 is graphed along the actual slope profile. The two years of the experiment are graphed separately.

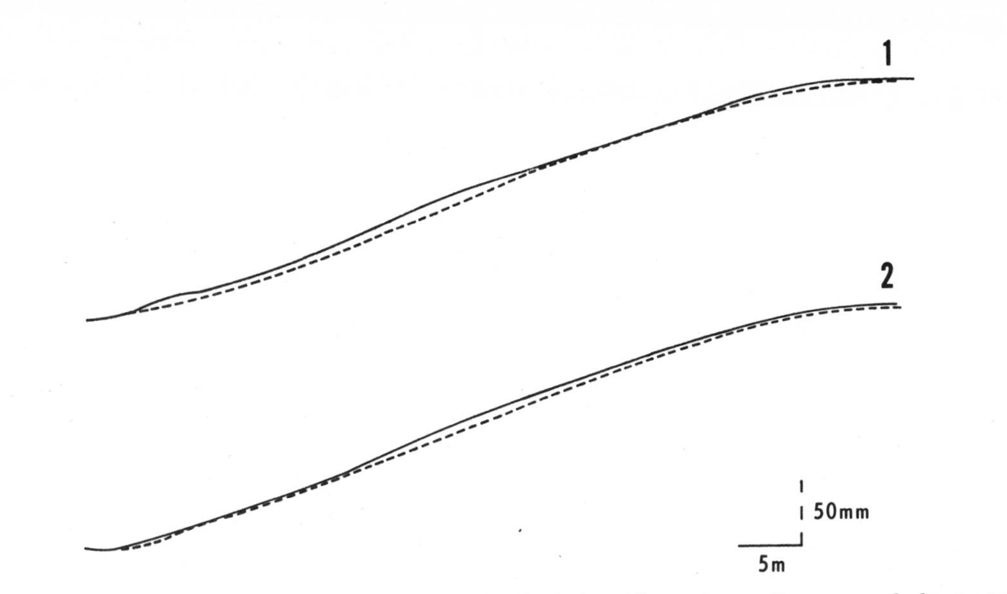

Fig. 19.3.--Waunafon W.: Ground retreat x 100 is graphed along the meter-unit measured slope profile. The two years of the experiment are graphed separately.

are graphed separately. In context, if the conditions of ground retreat on either of these slopes and in either year of the experiment were typical of the retreat to be experienced during the next one hundred years, then one of these four graphs is an accurate representation of future slope evolution. Happily the comparison of such graphs from Milfraen, Waunafon, and all the other instrumented slopes around Blaenavon, demonstrate that, for the most part, ground retreat during this experiment has proved to be more variable in space than in time.

Of course, mere multiplication is a very crude way of extrapolating future slope development from such data. Just suppose that these young irregular slopes are gradually evolving towards a more perfect state--call it an equilibrium form; call it "grade"--and that irregularities in the data disguise an underlying trend. Figure 19.4 displays time-trend surface maps for data from Milfraen S.W. (and represents an attempt to examine such an assertion). The vertical axis on these maps represents linear distance upslope, the horizontal axis represents the passage of time from June 1972, when the experiments were established, through to September 1974. The symbols on the map simply represent the migration of areas suffering variously: ground advance--A, ground retreat--R, or neutrality--N. (Recording error for this experiment was such that the standard deviation of original from repeated data records was c. 0.7 mm. The neutral slopes on this graph, therefore, were taken to include areas where there were recorded ground advances or ground retreats of less than 1.0 mms.) Unfortunately, the results of this analysis are far from satisfactory. The trend surface model fails to explain a reasonably high percentage of the data's variation. The goodness of fit of the Milfraen S.W. time-trend maps ranges from 0.14 for the first order surface to 0.41 for the quadratic surface. In fact, comparable explanation can be achieved by generating maps from random patterns (Tinker - 1969). Moreover, these maps display such edge effects that they should be useless prediction purposes. Nevertheless, some of these problems might be avoidable in analysis of data with a longer time span, and so possibly here, there may be the basis for a better method of predictive slope modelling from direct measurement data.

Direct measurement, however, must remain very close to the heart of the study of slope evolution. At least, here, one is dealing strictly with realities, and there is always the compensation that it is the very minutiae and variations which are revealed by, and conspire to complicate, direct measurement studies which are precisely the phenomena which future would-be slope model-

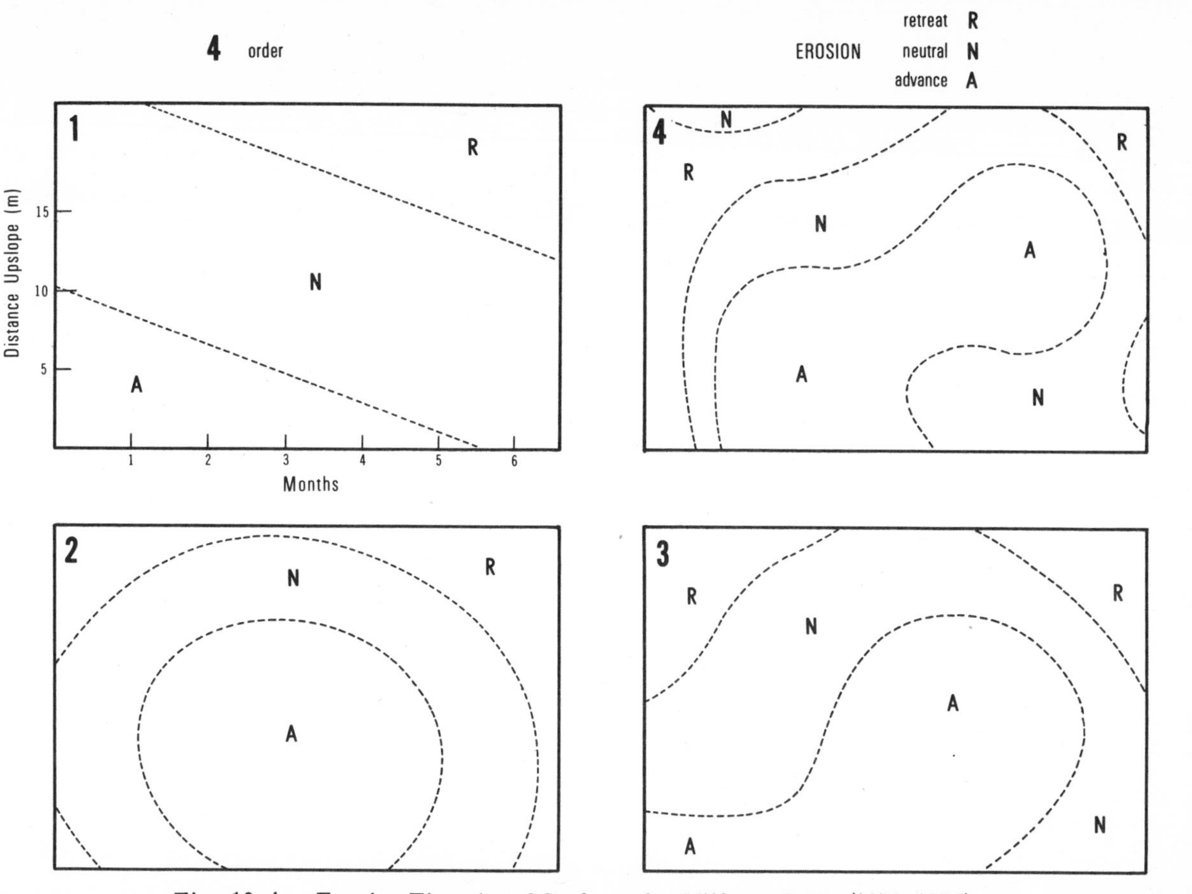

Fig. 19.4. --Erosion Time-trend Surfaces for Milfraen S. W. (1972-1974)

lers will have to account for and explain. This assurance is re-inforced by the observation that through two decades of studies involving the direct measurement of morphometric change there has been a surprising internal consistency in the measurements recorded and the fact that the broad conclusions drawn from similar studies in comparable locations have remained very much the same.

PROSPECTUS AND POSTSCRIPT

This monograph has been concerned with the results of a two year investigation into the nature of ground retreat and slope development on the artificial landforms of the Blaenavon area. This investigation is by no means ended. A third year's data has already been collected and awaits analysis, and it is hoped to continue data collection, albeit on a less regular basis, for as long as the landforms remain un-reclaimed and the erosion pins avoid the attention of vandals.

Several further developments have been initiated. A new erosion grid has been established on a plateau spoil tip in North Staffordshire. Retreat records from this site for the period February 1975-February 1976 are being examined in the light of the detailed climatological information collected at the nearby Keele Climatological Station. It is hoped that this study, which is being undertaken in collaboration with a former colleague at Keele University, will clear up some of the question marks hanging over the data from Milfraen and its relation to climatic phenomena. Further, two new erosion grids have been established on unvegetated strip mine spoil bank slopes at Henryetta, Eastern Oklahoma. These grids are located on slopes of identical morphology and construction but of different age. In associated studies, the relationships between slope age, morphology, and surface particle size are being examined. These studies are being undertaken in association with Dr. J. Goodman, University of Oklahoma, and at the Geomorphological Laboratory of the University of Chicago.

BIBLIOGRAPHY

Ahnert, F. (1970) A comparison of theoretical slope models with slopes in the field. Zeitschrift für Geomorphologie 9, 88-101.

Ahnert, F. (1971) A general and comprehensive theoretical model of slope profile development. University of Maryland, Occasional Papers in Geography 1. 95pp.

Anderson, D. H. and F. Bisal (1969) Snow cover effect on the erodible soil fraction. Canadian Journal of Soil Science 49, 287-96.

Anon. (1967) An analysis of mining activities. Minerals Magazine 117, 160-77.

Bakker, J. P. (1956) Niederländische Schriften über Hangentwicklung. Premier Rapport de la Commission pour l'Etude des Versants, Union Géographique Internationale, Amsterdam, 56-65.

Bakker, J. P. and J. W. N. Le Heux (1946) Projective-geometric treatment of D. Lehmann's theory of the transformation of steep mountain slopes. Koninklijke Nederlandsche Akademie van Wetenschappen, Series B, 49, 533-47.

Bakker, J. P. and J. W. N. Le Heux (1947) Theory on central recti-linear recession of slopes. Koninklijke Nederlandsche Akademie van Wetenschappen, Series B, 50, 959-66, 1154-62.

Balogh, J. and I. Matrai (1968) Effect of human activity on the catchment area. Hydrological Methods for Developing Water Resources Management (Budapest) 14, 79-201.

Barthel, H., Hornig, A. and L. Zapletal (1964) The principles of anthropogenic geomorphology. Abstracts of Papers, XXth Congress, International Geographical Union (London), 140.

Baver, L. D. (1938) Ewald Wollny--a pioneer in soil and water conservation research. Proceedings, Soil Science Society of America 3, 330-33.

Baver, L. D. (1956) Soil Physics. (3rd ed.) New York. 489pp.

Beaty, C. B. (1959) Slope retreat by gullying. Geological Society of America, Bulletin 70(2), 1479-82.

Beaumont, P. (ed.) (1973) Current research in geomorphology: British Geomorphological Research Group. Geo Abstracts, Norwich, 1-87.

Best, A. C. (1950) The size distribution of raindrops. Quarterly Journal of the Royal Meteorological Society 76, 302.

Bishop, A. W. (1973) The stability of tips and spoil heaps. Quarterly Journal of Engineering Geology 6, 335-76.

Blong, R. J. (1970) Development of discontinuous gullies in a pumice catchment. American Journal of Science 268, 369-83.

Bondarčuk, V. G. (1949) Antropogennyia formy reliefa. Voprosy geomorfologi, Moskva.

Bowden, K. L. (1961) A bibliography of strip mine reclamation, 1953-60. Department of Conservation, University of Michigan. 13pp.

Braithwaite, R. (1953) Scientific Explanation. Cambridge.

Brealey, S. C. (1966) Opencast mining. Min. J. Ann. Rev. May 1966, 143-53.

Brealey, S. C. and T. Atkinson (1968) Opencast mining. The Mining Engineer, December 1968, 147-63.

Brice, J. C. (1958) Origins of steps on loess mantled slopes. United States Geological Survey, Bulletin 1071C, 65-85.

Bridges, E. M. (1961) Aspect and time in soil formation. Agriculture 68(7), 360-63.

Bridges, E. M. (1969) Eroded soils of the Lower Swansea Valley. Journal of Soil Science 20(2), 236-45.

Bridges, E. M. and D. M. Harding (1971) Micro-erosion processes and factors affecting slope development in the Lower Swansea Valley. Institute of British Geographers, Special Publication 3, 65-80.

Brierly, J. K. (1956) Some preliminary observations on the ecology of pit heaps. Journal of Ecology 44, 381-89.

Brown, G. W. (1962) Piping erosion in Colorado. Journal of Soil and Water Conservation 17(5), 220-22.

Busch, D. C., Rochester, E. W. and C. L. Jernigan (1973) Soil crusting related to sprinkler intensity. Transactions, American Society of Agricultural Engineers 16, 808-9.

Butzer, K. W. (1973) Pluralism in geomorphology. Proceedings, Association of American Geographers 4, 39-43.

Campbell, I. A. (1970) Erosion rates in the Steveville Badlands, Alberta. Canadian Geographer 14, 202-16.

Campbell, I. A. (1970) Micro-relief measurements on unvegetated shale slopes. Professional Geographer 22(4), 215-21.

Campbell, I. A. (1973) Accelerated erosion in badland environments. Proceedings of the Ninth Canadian Hydrology Symposium, 18-29.

Campbell, I. A. (1974) Measurements of erosion on badland surfaces. Zeitschrift für Geomorphologie Suppl. 21, 122-37.

Carson, M. A. and M. J. Kirkby (1972) Hillslope: form and process. Cambridge Geographical Studies 3. 475pp.

Chorley, R. J. (1964) The nodal position and anomalous character of slope studies in geomorphological research. Geographical Journal 130, 70-73.

Chow, V. T. (1959) Open channel hydraulics. New York. 680pp.

Chwastek, J. (1970) Wpływ czynnikow gorniczo-geoloigicznych na formy zwałowisk. Czasopismo Geograficzne 4, 409-25.

Clayton, L. and J. R. Tinker (1971) Rates of hillslope lowering in the badlands of North Dakota. North Dakota University Water Resources Research Institute, Report WI-221-012-71. W73.09121. N.T.I.S. PB 220 355. 36pp.

Coates, D. R. (1972) Environmental geomorphology and landscape conservation. Dowden, Hutchinson, & Ross, Stroudsberg, Pa. Volume 1: Prior to 1900, 485pp (1972); volume 2: Urban, 480pp (1975); volume 3: Non-urban, 483pp (1973).

Colbert, E. C. (1956) Rates of erosion in the Chinle formation. Plateau 28(4), 73-76.

Cooke, R. U. and J. C. Doornkamp (1974) Geomorphology in environmental management: an introduction. Oxford. 413pp.

Cooke, R. U. and R. W. Reeves (1976) Arroyos and environmental change in the American Southwest. Oxford. 213pp.

Cornwell, S. M. (1971) Anthracite mining spoils in Pennsylvania. Journal of Applied Ecology 8(2), 401-9.

Crabtree, K. (1971) Overton Down Experimental Earthwork, Wiltshire 1968: Geomorphology of the ditch section. Proceedings, Bristol University Spelaeological Society 12(3), 237-44.

Dauksza, L. and A. Kotarba (1973) An analysis of the influence of fluvial erosion in the development of a landslide slope (using the application of queueing theory). Studia Geomorphologica Carpatho-Balcanica 7, 91-104.

Davis, W. M. (1898) The grading of mountain slopes. Science 7, 1449.

Davis, W. M. (1899) The geographical cycle. Geographical Journal 14, 481-504.

Davison, S. (1974) Unpublished paper presented to the Annual Meeting of the British Association, Stirling.

Demek, J. (1973) Quaternary relief development and man. Geoforum 15, 68-71.

Denison, W. J. (1959) Pit heap reclamation. British Housing and Planning Review 14, 18-21.

De Ploey, J. (1969) L'érosion pluviale: expériences à l'aide de sables traceurs et bilans morphogéniques. Acta Geographica Lovaniensia 7, 1-28.

De Ploey, J. (1972) A quantitative comparison between rainfall erosion capacity in a tropical and middle-latitude region. Geographica Polonica 23, 138-50.

De Ploey, J. (1974) Mechanical properties of hillslopes and their relation to gullying in central semi-arid Tunisia. Zeitschrift für Geomorphologie, Suppl. 21, 177-90.

Devdarijani, A. S. (1954) Anthropogenniya formy reliefa. Voprosy geografigeomorfologi, Moskva, 117-20.

Dietrich, T. L. and J. R. Meiman (1974) Hydrologic effects of patch cutting of lodgepole pine. Colorado State University, Hydrology Papers 66. 31pp.

Diseker, E. G. and J. T. McGinnis (1967) Evaluation of climatic, slope and site factors on erosion from unprotected roadbanks. Transactions of the American Society of Agricultural Engineers 10, 9-11 and 14.

Diseker, E. G. and E. C. Richardson (1961) Roadside sediment production and control. Transactions of the American Society of Agricultural Engineers 4, 62-67.

Diseker, E. G. and E. C. Richardson (1962) Erosion rates and control methods on highway cuts. Transactions of the American Society of Agricultural Engineers 5, 153-55.

Diseker, E. G. and J. M. Sheridan (1971) Predicting sediment yield from roadbanks. Transactions of the American Society of Agricultural Engineers 14, 102-5.

Diseker, E. G. and R. E. Yoder (1936) Sheet erosion studies on Cecil Clay. Alabama Agricultural Experimental Station, Bulletin 245.

Disrud, L. A. and R. K. Krauss (1971) Examining the process of soil detachment from clods exposed to wind-driven simulated rainfall. Transactions, American Society of Agricultural Engineers 14, 90-92.

Doubleday, G. P. (1969) The assessment of colliery spoil as a soil forming material. Proceedings, North of England Soils Discussion Group 6, 5-13.

Doubleday, G. P. (1972) Development and management of soils on pit heaps. Landscape Reclamation 2, I. P. C., Newcastle, 25-35.

Down, C. G. (1973) Life form succession in plant communities on colliery waste tips. Environmental Pollution 5(1), 19-22.

Downes, R. G. (1946) Conservation problems on solodic soils in the State of Victoria (Australia). Journal of Soil and Water Conservation 1(5), 228-32.

Dragoun, F. J. (1969) Effects of cultivation and grass on surface runoff. Water Resources Research 5(5), 1078-83.

Ellison, W. D. (1944) Studies of raindrop erosion. Agricultural Engineering 25, 131-36, 181-82.

Ellison, W. D. (1947) Soil erosion studies I-VII. Agricultural Engineering 28, 145-46, 197-201, 245-48, 297-300, 349-51, 402-5, 442-44.

Ellison, W. D. (1952) Raindrop energy and soil erosion. Empire Journal of Experimental Agriculture 20, 81-97.

Emmett, W. W. (1965) The Vigil Network: methods of measurement and a sampling of data collected. International Association for Scientific Hydrology, Publication 66, 89-106.

Emmett, W. W. (1970) The hydraulics of overland flow. United States Geological Survey Professional Paper 662-A. 68pp.

Emmett, W. W. (1974) Channel aggradation in the western United States. Zeitschrift für Geomorphologie Suppl. 21, 52-62.

Engelen, G. N. (1973) Runoff processes and slope development in badlands, National Monument, South Dakota. Journal of Hydrology 18, 55-79.

Epstein, E. and W. J. Grant (1967) Soil losses and crust formation as related to some soil physical properties. Proceedings, Soil Science Society of America 31, 547-50.

Evans, G. R. (1973) Hutton's shearwaters initiating local soil erosion in the seaward Kaikowa Range (N. Z.). New Zealand Journal of Science 16(3), 637-42.

Evans, R. (1967) On the use of welding rod for erosion and deposition pins/ modified depth gauge for erosion rod measurement. Revue de Géomorphologie Dynamique 17(4), 165.

Everett, D. H. (1961) The thermodynamics of frost damage to porous solids. Transactions, Faraday Society 57, 1541-51.

Fair, T. J. (1947) Slope form and development in the interior of Natal. Transactions, Geological Society of South Africa 50, 105-20.

Fair, T. J. (1948) Slope form and development in the coastal hinterland of Natal. Transactions, Geological Society of South Africa 51, 37-53.

Farmer, E. E. (1973) Relative detachability of soil particles by simulated rainfall. Proceedings, Soil Science Society of America 37, 629-33.

Faulkner, H. (1974) An allometric growth model for competitive gullies. Zeitschrift für Geomorphologie Suppl. 21, 76-87.

Fels, E. (1934) Der Mensch als Gestalter der Erdoberfläche. Petermanns Geographische Mitteilungen.

Fels, E. (1935) Der Mensch als Gestalter der Erde. Leipzig.

Fels, E. (1957) Anthropogene Geomorphologie. Scientia (Bologna) 51.

Fels, E. (1965) Nochmals: Anthropogene Geomorphologie. Petermanns Geographische Mitteilungen 109(1), 9-15.

Fisher, O. (1866) On the disintegration of a chalk cliff. Geological Magazine 3, 354-56.

Fitton, A. H., Gibbons, J., Precious, G., Webber, J. and G. H. Theobald (1959) Experiments on rehabilitation of opencast sites. Experimental Husbandry 4, 58-73.

Foster, R. L. and G. L. Martin (1969) Effect of unit weight and slope on erosion. Proceedings of the American Society of Civil Engineers, Journal of the Irrigation and Drainage Division IR4, 551-61.

Free, G. R. (1952) Soil movement by raindrops. Agricultural Engineering 33, 491-94, 496.

Fuller, M. L. (1922) Some unusual erosion features in the loess of China. Geographical Review 12(4), 570-84.

Gerlach, T. (1966) Wspotczesny rowoj stokow w dorzeczu gornegu Grajcarka. Prace Geograficzne T. G. PAN 52, 111 pp.

Gerrard, A. J. W. and D. A. Robinson (1971) Variability in slope measurements. Transactions, Institute of British Geographers 54, 45-54.

Gil, E. (1974) An attempt to determine the size of washing in the Bystrzanka catchment basin. Studia Geomorphologica Carpatho-Balcanica 8, 105-14.

Gil, E. and J. Słupik (1972) The influence of the plant cover and land use on the surface run-off and wash down during heavy rain. Studia Geomorphologica Carpatho-Balcanica 6, 181-90.

Golomb, B. (1964) Anthropogeomorphology: the study of man-made landforms. Abstracts of Papers, XXth Congress, International Geographical Union (London) 168.

Guney, M. (1968) Oxidation and spontaneous combustion of coal--review of individual factors. Colliery Guardian January 1968, 105-10, February 1968, 137-43.

Hadley, R. F. (1965) Selecting sites for the observation of geomorphic and hydrological processes through time. International Association for Scientific Hydrology, Publication 66, 217-33.

Hadley, R. F. and G. C. Lusby (1967) Runoff and hillslope erosion resulting from a high-intensity thunderstorm near Mack, Western Colorado. Water Resources Research 3(1), 139-48.

Hadley, R. F. and B. N. Rolfe (1955) Development and significance of seepage steps in slope erosion. Transactions, American Geophysical Union 36(5), 792-804.

Hadley, R. F. and S. A. Schumm (1961) Sediment sources and drainage basin characteristics in the Upper Cheyenne River Basin. United States Geological Survey Water Supply Paper 1531B, 135-98.

Hall, I. G. (1957) The ecology of disused pit heaps in England. Journal of Ecology 45, 690-720.

Haigh, M. J. (1973) The zonation of asymmetric valleys in North America and Northern Asia. Northern Universities Geographical Journal 10, 44-55.

Haigh, M. J. (1974a) Ground retreat and slope development on colliery spoil mounds and infilled opencast sites near Blaenavon (Mon.). Unpublished paper presented to the Annual Conference of the Institute of British Geographers, Norwich.

Haigh, M. J. (1974b) Ground retreat and slope development on colliery spoil mounds near Blaenavon (Gwent). Unpublished paper presented to "Soils, Geomorphology and Engineering," a Symposium of the British Geomorphological Research Group, Portsmouth.

Haigh, M. J. (1974c) Problems of direct measurement in studies of hill-slope evolution. Unpublished paper presented to the British Geomorphological Research Group's Small Research Group on Geomorphometry's Symposium: "Slope Morphometry and Models of Slope Evolution," Birmingham.

Haigh, M. J. (1975) National preferences in citation selection: British geographical geomorphology, 1970-1973. Area 7(3), in press.

Haigh, M. J. (1977) Erosion of surface-mine disturbed land at Henryetta Oklahoma. Proceedings, Oklahoma Academy of Science 57, 173-75.

Harvey, A. M. (1974) Gully erosion and sediment yield in the Howgill Fells, Westmorland. Institute of British Geographers, Special Publication 6, 59-72.

Hashimoto, K. (1974) A theory on the mechanism of an outbreak of spontaneous combustion of coal. Memoirs, Faculty of Engineering Hokkaido University 13(4), 271-76.

Heede, B. (1967) The fusion of discontinuous gullies. Bulletin, International Association for Scientific Hydrology 12(4), 42-50.

Heede, B. (1970) Morphology of gullies in the Colorado Rocky Mountains. Bulletin, International Association for Scientific Hydrology 15(2), 79-89.

Heede, B. H. (1974) Stages of development of gullies in western United States of America. Zeitschrift für Geomorphologie 18(3), 260-71.

Hirano, M. (1968) A mathematical model of slope development. Journal of Geoscience (Osaka City University) 11, 13-52.

Hoekstra, P. (1967) Moisture movement to a freezing front. International Association for Scientific Hydrology, Publication 78.

Holy, M. and K. Vrana (1970) The influence of the vegetative cover on the texture of the top-soil layer during erosion. Proceedings, International Water Erosion Symposium (Czechoslovak National Committee of the International Commission on Irrigation and Drainage, Praha) 2, 183-91.

Hornig, A. (1963) The phenomena of anthropogenic changes of the earth surface in Upper Silesia. Annales Silesiae 2, 81-96.

Horton, R. E. (1945) Erosional development of streams and their drainage basins: hydrophysical approach to quantitative morphology. Geological Society of America, Bulletin 56, 275-370.

Imeson, A. C. (1971) Heather burning and soil erosion on the North Yorkshire Moors. Journal of Applied Ecology 8, 537-42.

Imeson, A. C. (1974) The origin of sediment in a moorland catchment with particular reference to the role of vegetation. Institute of British Geographers, Special Publication 6, 59-72.

Iogansen, N. K. (1970) Klassifikatsiya antropogennykh landshaftov. Vestnik Leningradskogo Universiteta Geologiya-Geographica 24(4), 57-62.

Jewell, P. A. (ed.) (1963) The experimental earthwork on Overton Down, Wiltshire 1960. British Association for the Advancement of Science, Research Committee Report. 100pp.

Jones, G. A., Penman, A. D. N. and C. R. V. Tandy (1972). Industrial spoil tips. Civil Engineering Problems of the South Wales Valleys, Institution Civil Engineers, London, 81-91, 92-105.

Jones, N. O. (1968) The development of piping erosion. Unpublished Ph.D. thesis, Department of Geosciences, University of Arizona. 163pp.

Jones, P. E. (1974) The iron industry and the changing cultural landscape of the Eastern Valley of Monmouthshire. Unpublished B.Ed. dissertation, Geography Department, Caerleon College of Education, University of Wales. 47pp.

Karrasch, H. (1970) Das Phänomen der Klimabedingten Reliefasymmetrie in Mitteleuropa. Göttinger Geographische Abhandlungen 56. 299pp.

Karrasch, H. (1972) The planetary and hypsometric variation of valley asymmetry. International Geography 1, 30-33. (International Geographical Union Congress, Montreal).

Kennard, M. F., Knill, J. L. and P. R. Vaugham (1967) The geotechnical properties of the Carboniferous shale at Balderhead Dam. Quarterly Journal of Engineering Geology 1, 3-24.

Kennedy, B. A. (1973) Valley asymmetry and the relative importance of stream processes in selected areas of North America. Unpublished paper. Annual conference of the Institute of British Geographers, Birmingham, 1973.

Kilinc, M. and E. V. Richardson (1973) Mechanics of soil erosion from overland flow generated by simulated rainfall. Colorado State University, Hydrology Paper 63. 54pp.

Kirkby, A. V. T. and M. J. Kirkby (1974) Surface wash at the semi-arid break of slope. Zeitschrift für Geomorphologie Supplementband 21, 151-76.

Kowobari, T. S., Rice, C. E. and J. E. Garton (1972) Effect of roughness elements on hydraulic resistance for overland flow. Transactions, American Society of Agricultural Engineers 15(5), 979-84.

Laws, J. O. (1940) Recent studies in rain drops and erosion. Agricultural Engineering 2, 431-34.

Legget, R. F. (1969) Man as a geological agent. Byulleten' Moskovskogo Obshchestva Ispytateleley prirody, Odtel geologicheskiy, 56-64.

Lehmann, O. (1933) Morphologische Theorie der Verwitterung von Steinschlagwänden. Vierteljahrsschrift Naturforsch. Gesellschaft, Zurich 78, 83-126.

Leliavsky, S. (1955) An introduction to fluvial hydraulics. London. 257pp.

Lemos, P. and J. F. Lutz (1957) Soil crusting and some factors affecting it. Proceedings, Soil Science Society of America 21, 485-91.

Leopold, L. B. and W. W. Emmett (1972) Some rates of geomorphological process. Geographia Polonica 23, 27-35.

Leopold, L. B., Emmett, W. W. and R. M. Myrick (1966) Channel and hillslope processes in a semi-arid area of New Mexico. United States Geological Survey Professional Paper 352-G, 193-253.

Leopold, L. B. and J. P. Miller (1956) Ephemeral streams--hydraulic factors and their relation to the drainage net. United States Geological Survey Professional Paper 282-A. 37pp.

Leopold, L. B., Wolman, M. G. and J. P. Miller (1964) Fluvial processes in geomorphology. San Francisco. 522pp.

Li, R-M. and H. W. Shen (1973) Effect of tall vegetation on flow and sediment. Proceedings of the American Society of Civil Engineers: Journal of the Hydraulics Division HY5, 793-814.

Lixandru, Ch. (1968) Contributii la stabilirea distantei critice de eroziune. Stiinta Solului 6(1), 12-18.

Lloyd J. (1906) Old South Wales iron-works (cited by Jones - 1974).

Louis, H. (1968) Allgemeine geomorphologie. Berlin.

Lowenthal, D. (1965) Introduction to G. P. Marsh (1864) Man and Nature. Harvard. 656pp.

Luke, J. C. (1972) Mathematical models for slope evolution. Journal of Geophysical Research 77(14), 2460-64.

Marosi, S. (1971) Some questions of anthropogenic slope development. Acta Geographica Debrecina 10, 133-42.

Marosi, S. and J. Szilard (1969) A lejtöfejlödes nehany kerdese a talajkepzödes es a talajpusztulas tükreben. Földrajzi Ertesitö 18(1), 53-68.

Marsh, G. P. (1864) Man and Nature. New York. 650pp.

McIntyre, D. S. (1958a) Permeability measurements of soil crusts formed by raindrop impact. Soil Science 85, 186-89.

McIntyre, D. S. (1958b) Soil splash and the formation of surface crusts by raindrop impact. Soil Science 85, 261-66.

McKecknie Thompson, G. and S. Rodin (1972) Colliery spoil tips--after Aberfan. Institution Civil Engineers, London. 60pp.

McKechnie Thompson, G. and S. Rodin (1973) Colliery spoil tips after Aberfan (Discussion). Proceedings, Institution Civil Engineers 55(2), 677-712.

Matthews, B. (1962) Frost-heave cycles at Schefferville. McGill Sub-arctic Research Report 12, 112-25.

Melton, M. A. (1957) An analysis of the relations among the elements of climate, surface properties, and geomorphology. Columbia University, Department of Geology, Technical Report 11. 102pp.

Meyer, L. D. and L. A. Kramer (1968) Relation between land-slope shape and soil erosion. Unpublished paper presented to the Winter Meeting of the American Society of Agricultural Engineers, Chicago. A.S.A.E. Paper 68-749.

Meyer, L. D. and L. A. Kramer (1969) Erosion equations predict land slope development. Agricultural Engineering 50, 522-23.

Meyer, L. D., Wischmeier, W. H. and W. H. Daniel (1971) Erosion, runoff and revegetation of denuded construction sites. Transactions, American Society of Agricultural Engineers 14, 138-41.

Mihara, Y. (1951) Raindrop and soil erosion. Bulletin, National Institute of Agricultural Science (Nishigahara, Tokyo), Series A, 1. 51pp.

Miller, J. P. and L. B. Leopold (1962) Simple measurement of morphological changes in river channels and hill slopes. Arid Zone Research 20, 421-27.

Mizutani, T. (1969) Erosional processes of youthfully dissected strato-volcanos in Japan. Geographical Reports of Tokyo Metropolitan University 4, 11-22.

Mizutani, T. (1970) Quantitative analysis on the process of slope development. Geographical Reports of Tokyo Metropolitan University 5, 49-60.

Molyneux, J. K. (1963) Some ecological aspects of colliery waste tips around Wigan, South Lancashire. Journal of Ecology 51, 315-21.

Monmouthshire Derelict Land Joint Committee (1971) Comprehensive report on derelict land. M.D.L.J.C. 66pp.

Morgan, G. S. (1965) South Wales Coalfield--Second Century, retrospective and prospect. The Mining Engineer February 1965, 278-83.

Morisawa, M. E. (1964) Development of drainage systems on an upraised lake floor. American Journal of Science 262, 340-54.

Morris, L. (1974) The accuracy of slope measurements using the slope pantometer. Unpublished paper presented to the British Geomorphological Research Group's Small Research Group on Geomorphometry's Symposium: "Slope Morphometry and Models of Slope Development"--Birmingham, November, 1974.

Murayama, I. (1964) Soil erosion by winds on the eastern foot of the Zao Volcanoes. Saito Ho-on Museum Research Bulletin 33, 9-16.

Musgrave, G. (1947) The quantitative evaluation of factors in water erosion. Journal of Soil and Water Conservation 2, 133-38.

National Coal Board (1950) Plan for coal. N.C.B., London.

Nir, D. and M. Klein (1974) Gully erosion induced in land use in a semi-arid terrain (Nahal Shiqma, Israel). Zeitschrift für Geomorphologie Suppl. 21, 191-201.

Palmer, M. E. (1969) Weathering reactions in colliery spoil. Proceedings, North of England Soils Discussion Group 6, 1-4.

Palmer, R. S. (1965) Waterdrop impact forces. Transactions, American Society of Agricultural Engineers 8(1), 69-72.

Panos, V. (1973) The development dynamics of small landscape forms in the weathering and vegetation mantles of the Belanské Tatry Mts. (Czechoslovakia). Acta Universitatis Palackianae Olomoucensis Facultas Rerum Naturalium 42, Geographica-Geologica 13, 109-26.

Pantastico, E. B. and T. I. Ashaye (1965) Demonstration of the effect of stone layer on soil transport and accretion. Experimental Pedology (ed.: Hallsworth, E. G. and D. V. Crawford). Butterworth, London. 384-90.

Parker, G. G. (1963) Piping, a geomorphic agent in landform development of the drylands. International Association for Scientific Hydrology, Publication 65, 103-13.

Penck, W. (1924) Die Morphologische Analyse. Stuttgart. (Translation: H. Czech and K. C. Boswell [1953] The morphological analysis of landforms. London.)

Pinczes, Z. (1971) Erosion forms and erosion control in the vinelands of the Tokaj Mountains. In Lacko, L. and P. Polyanszky (eds.), Abstracts of Papers from the International Geographical Union's Regional Conference, Budapest, 60.

Pitty, A. F. (1967) Some problems in selecting a ground-surface length for slope angle measurement. Revue de Géomorphologie Dynamique 18, 66-71.

Popov, I. V. (1968) Human activity as a geological factor. Thematic Collection of the All-Union Scientific Research Institute of Hydro-geology and Engineering Geology 2, 56-61.

Prohaska, J. A. (1974) Remarks on: the physical geographic problems associated with lignite strip mining in Western North Dakota. Great Plains and Rocky Mountain Geographical Journal, 122.

Rees, W. J. (1955) Some preliminary observations on the flora of derelict land. Proceedings, Birmingham Natural History and Philosophical Society 18, 119-29.

Repelewska, J. (1968) Procesy erozyjne na zwałach kopalnianych. Czasopismo Geograficzne 39(1), 31-43.

Richardson, J. A. (1958a) The effect of temperature on the growth of plants on pit heaps. Journal of Ecology 46(3), 537-46.

Richardson, J. A. (1958b) Derelict pit heaps and their vegetation. Planning Outlook 4, 15-22.

Richardson, J. A. and E. F. Greenwood (1967) Soil moisture tension in relation to plant colonisation of pit heaps. Proceedings, University of Newcastle-upon-Tyne Philosophical Society 1(9), 129-36.

Riedl, D. (1970) Theoretical bases for the design of anti-erosion measures. Proceedings, International Water Erosion Symposium (Czechoslovak National Committee of the International Commission on Irrigation and Drainage, Praha) 2, 1-12.

Rougerie, G. (1960) Le façonnement actuel des modèles en Côte d'Ivoire forestière. Mémoires, I.F.A.N. (Dakar) 58. 600pp.

Rowlinson, D. L. (1968) The effect of unit weight and rainfall intensity on the erosion of unprotected slopes. Unpublished (partial) M.Sc. thesis, Montana State University. (In Rowlinson and Martin - 1971.)

Rowlinson, D. L. and G. L. Martin (1971) Rational model describing slope erosion. Proceedings of the American Society of Civil Engineers, Journal of the Irrigation and Drainage Division IR1, 39-50.

Rushton, J. (1973) Methods of working in the American coal mining industry. The Mining Engineer February, 1973, 251-59.

Ryabchikov, A. M. (1971) Planetarnye izmeneniya prirodnoy sredy proizvodstvom. Vestnik Moskovoskii Gosudarstvennyi Universitat, Seriya Geografiya 5, 8-17.

Sabata, M. (1970) The influence of vegetation cover on water erosion. Proceedings, International Water Erosion Symposium (Czechoslovak National Committee of the International Commission on Irrigation and Drainage, Praha) 2, 173-83.

Savigear, R. A. G. (1952) Some observations on slope development in South Wales. Transactions, Institute of British Geographers 18, 31-51.

Savigear, R. A. G. (1956) Technique and terminology in the investigation of slope forms. Premier rapport de la Commission pour l'Etude des Versants, Union Géographique Internationale, Amsterdam.

Scheidegger, A. E. (1961a) Theoretical geomorphology. Springer, Berlin.

Scheidegger, A. E. (1961b) Mathematical models of slope development. Geological Society of America, Bulletin 72, 37-50.

Schramm, J. R. (1966) Plant colonisation studies of black wastes from anthracite mining in Pennsylvania. Transactions, American Philosophical Society 56, 1-194.

Schulter, U. (1973) Die Entwicklung von Heckenlagen auf saurem teriärem Abraummaterial im Braunkohlen revier Helmstedt nach secs Vegetationsperioden. Landschaft und Stadt 1, 42-48.

Schumm, S. A. (1956a) Evolution of drainage systems and slopes in badlands at Perth Amboy, New Jersey. Geological Society of America, Bulletin 67, 597-646.

Schumm, S. A. (1956b) The role of creep and rainwash on the retreat of badland slopes. American Journal of Science 254, 693-706.

Schumm, S. A. (1964) Seasonal variations in erosion rates on hill-slopes in western Colorado. Zeitschrift für Geomorphologie Supplementband 5, 215-38.

Schumm, S. A. (1967) Erosion measured by stakes. Revue de Géomorphologie Dynamique 17, 161-62.

Schumm, S. A. and G. C. Lusby (1963) Seasonal variation of infiltration capacity and runoff on hillslopes in Western Colorado. Journal of Geophysical Research 68, 3655-66.

Schumm, S. A. and M. P. Moseley (eds.) (1973) Slope morphology (Benchmark Papers in Geology). Stroudsburg, Pennsylvania. 454pp.

Selby, M. J. (1972) The relationships between land use and erosion in the Central North Island, New Zealand. Journal of Hydrology (N. Z.) 11(2), 73-87.

Selby, M. J. (1973) An investigation into causes of runoff from a catchment of pumice lithology in New Zealand. Hydrological Sciences, Bulletin 18(3), 255-80.

Selby, M. J. and P. J. Hosking (1973) The erodibility of pumice soils of the North Island, New Zealand. Journal of Hydrology (N. Z.) 12(1), 32-56.

Shchukin, I. S. (1973) Some thoughts on the origin, development, and retreat of slopes. Soviet Geography 14(7), 458-67.

Shen, H. W. and R-M. Li (1973) Rainfall effect on sheet flow over smooth surface. Proceedings of the American Society of Civil Engineers: Journal of the Hydraulics Division HY5, 771-92.

Sherlock, R. L. (1922) Man as a geological agent. London.

Sherlock, R. L. (1923) The influence of Man as an agent in geographical change. Geographical Journal 61, 258-73.

Sherlock, R. L. (1931) Man's influence on the earth. Home University Library, Butterworth, London. 256pp.

Smalley, I. J. (1972) Boundary conditions for flowslides in fine-particle mine waste tips. Transactions, Institute Mining Engineers 81, A31-A37.

Smith, D. D. and W. H. Wischmeier (1957) Factors affecting sheet and rill erosion. Transactions, American Geophysical Union 38(6), 889-96.

Smith, D. D. and W. H. Wischmeier (1962) Rainfall erosion. Advances in Agronomy 14, 109-48.

Smith, G. N. (1968) Coal spoil heaps and their stability. Ground Engineering 1, July, 11-13, 20; September, 42-45.

Smith, G. N. (1968) Coal spoil heaps--site investigations and stability analysis. Colliery Guardian 216, 585-86.

Smith, G. N. (1971) Elements of soil mechanics for civil engineers and mining engineers. Crosby Lockwood, London. 340pp.

Smith, R. M., Tryon, E. H. and E. H. Tyner (1971) Soil development on mine spoil. West Virginia University Agricultural Experiment Station, Bulletin 604T, 47pp.

Soons, M. J. (1968) Erosion by needle ice in the Southern Alps, New Zealand. Arctic and Alpine Environments, ed. H. Wright and Osburn. Indiana. 217-27.

Spears, D. A., Taylor, R. K. and R. Till (1971) A mineralogical investigation of a spoil heap at Yorkshire Main Colliery. Quarterly Journal of Engineering Geology 3, 239-52.

Stepanov, I. N. (1967) Asymmetrical development of soils on the northern and southern exposures of the Central Tien Shan. Soviet Soil Science 2, 170-76.

Streeter, D. (1975) Measurements of erosion on chalk downland footpaths. Unpublished contribution to: Annual Conference, Institute of British Geographers, University of Oxford.

Tackett, J. L. and R. W. Pearson (1965) Some characteristics of soil crusts formed by simulated rainfall. Soil Science 99, 407-13.

Taylor, R. K. (1974) Colliery spoil heap materials--time dependent changes. Ground Engineering July 1974, 24-27.

Taylor, R. K. and D. A. Spears (1972) The geotechnical characteristics of a spoil heap at Yorkshire Main Colliery. Quarterly Journal of Engineering Geology 5, 243-63.

Temple, P. H. and D. H. Murray-Rust (1972) Sheetwash measurements on erosion plots at Mfumbwe, Eastern Uluguru Mountains, Tanzania. Geografiska Annaler 54A(3-4), 195-202.

Thomas, T. M. (1965) Sheet erosion induced by sheep in the Pumlumon (Plylimon) area, Mid-Wales. Rates of Erosion and Weathering in the British Isles (Bristol/B. G. R. G.), 11-14.

Thomas, T. M. (1966) Derelict land in South Wales. Town Planning Review 37(2), 125-41.

Thompson, D. N. and R. J. Hutnik (1971) Environmental characteristics affecting plant growth on deep-mine coal refuse banks. Pennsylvania State University, University Park, College of Earth and Mineral Sciences, Special Research Report SR-88, 81pp.

Thornes, J. B. (1971) State, environment and attribute in scree-slope studies. Institute of British Geographers, Special Publication 3, 49-63.

Timm, H., J. C. Bishop et al. (1971) Soil crusting effects on potato plant emergence and growth. California Agriculture 25(8), 5-7.

Tinker, J. R., Jr. (1970) Rates of hillslope lowering in the badlands of North Dakota. Unpublished Ph. D. thesis, University of North Dakota.

Tinkler, K. J. (1969) Trend surfaces with low "explanations"; the assessment of their significance. American Journal of Science 267, 114-23.

Tivy, J. (1962) An investigation of certain slope deposits in the Lowther Hills, Southern Uplands of Scotland. Transactions, Institute of British Geographers 30, 59-73.

Tuckfield, C. G. (1964) Gully erosion in the New Forest, Hampshire. American Journal of Science 262, 795-807.

Tuckfield, C. G. (1965) Rate of erosion by gullying in the New Forest. Rates of Erosion and Weathering in the British Isles, (British Geomorphological Research Group, Bristol), 15-19.

U.S. Department of the Interior (1967) Surface mining and our environment. Washington. 124pp.

Vermes, J. (1971) A pluviacio folyamatanak es formakepzesenek vizsgalata. Földrajzi Ertesitö 20(4), 365-82.

Wali, M. K. (ed.) (1973) Some environmental aspects of strip mining in North Dakota. North Dakota Geological Survey Educational Series 5. 85pp.

Warkentin, B. P., Bolt, G. H. and R. D. Miller (1957) Swelling pressure of Montmorillonite. Proceedings, Soil Science Society of America 21(5), 495-97.

Warwick, J. (1958) Plant nutrition on colliery waste. Unpublished Ph. D. thesis, University of Birmingham.

White, D. A. (1963) Gully erosion in Southern England, a study of contemporary erosion with particular reference to gully erosion in the heathlands of S. England in relation to geomorphology, ecology and land-use. Unpublished M. Sc. thesis, University of London. 199pp.

Whyte, R. O. and J. W. B. Sisam (1948) The establishment of vegetation on industrial wasteland. Commonwealth Agricultural Bureaux Joint Publication 48. 78pp.

Williams, M. A. J. (1973) The efficacy of creep and slope-wash in tropical and temperate Australia. Australian Geographical Studies 9, 62-78.

Wischmeier, W. H. (1970) Relation of soil erosion to crop and soil management. Proceedings, International Water Erosion Symposium (Czechoslovak National Committee of the International Commission on Irrigation and Drainage, Praha) 1, 201-19.

Wischmeier, W. H. and D. D. Smith (1958) Rainfall energy and its relation to soil loss. Transactions, American Geophysical Union 39, 285-91.

Woldenberg, M. J. (1966) Horton's Laws justified in terms of allometric growth and steady state in open systems. Geological Society of America, Bulletin 77, 431-34.

Woodburn, R. and J. Kozachyn (1956) A study of the relative erodibility of a group of Mississippi gully soils. Transactions, American Geophysical Union 37(6), 749-53.

Yatsu, E. (1966) Rock control and geomorphology. Tokyo.

Yatsu, E. (1971) Landform material science, rock control in geomorphology. Proceedings, 1st Guelph Symposium on Geomorphology. University of Guelph Geographical Publication 1, 49-73.

Young, A. (1960) Soil movement by denudational processes on slopes. Nature 188, 120-22.

Young, A. (1964) Slope profile analysis. Zeitschrift für Geomorphologie Supplementband 5, 17-24.

Young, A. (1973) Slopes. Oliver & Boyd Geomorphology Text 3, Edinburgh. 288pp.

Young, A. (1974) The rate of slope retreat. Institute of British Geographers, Special Publication 7, 65-78.

Young, R. A. and D. K. Mutchler (1967) Effect of slope shape on erosion and runoff. Unpublished paper presented at the Winter Meeting of the American Society of Agricultural Engineers, Detroit. A.S.A.E. Paper 67-706.

Young, R. A. and D. K. Mutchler (1969) Effect of slope shape on erosion and runoff. Transactions, American Society of Agricultural Engineers 12, 231-33, 239.

Young, R. A. and D. K. Mutchler (1969) Soil movement on irregular slopes. Water Resources Research 5(5), 1084-89.

Zaborski, B. (1972) On the origin of gullies in loess. Acta Geographica Debrecina 10, 109-11.

Zapletal, L. (1960) Anthropogenous forms of relief. Abstracts of Papers, XIXth International Geographical Union Congress (Stockholm), 323-24.

Zapletal, L. (1964) The anthropogenic factor in geographical environment and population. Abstracts of Papers, XXth Congress, International Geographical Union (London), 292.

Zapletal, L. (1968) Genetico-morphologika klasifikace anthropogennich forem reliefu. Acta Universitatis Palackianae Olomoucensis, Facultas Rerum Naturalium 23, Geographica-Geologica 8, 239-427.

Zapletal, L. (1969) Uvod do anthropogenni geomorfologie. Ucebni texty vysokych skol. Olomouc. 278pp.

Zapletal, L. (1973a) Antropogenni relief strdoasijskeho podhuri velehor zailijskij Alatau. Acta Universitatis Palackianae Olomoucensis, Facultas Rerum Naturalium 42, Geographica-Geologica 13, 197-222.

Zapletal, L. (1973b) Neprime antropogenni geomorfologicke procesy a jejicj vliv na zemsky povrch. Acta Universitatis Palackianae Olomoucensis, Facultas Rerum Naturalium 42, Geographica-Geologica 13, 239-61.

Zingg, A. (1940) Degree and length of landslope as it affects soil loss in runoff. Agricultural Engineering 21, 59-64.

ABSTRACT

The evolution of slopes and gullies developing on colliery spoil mounds and opencast-mine fill has been monitored by means of erosion pins for a period of two years. Profile development on colliery spoil mounds created by tipping and compaction proceeded by slope foot extension and accumulation, the parallel retreat of an abbreviating main-slope segment, and the rapid amelioration of the upper convexity. Vegetated slopes retreated at 3.1 mms p.a. Retreat was more rapid on south and west facing slopes than those facing north and east. Unvegetated slopes retreated at 4.6 mms p.a. The devegetation of these mounds was usually initiated at the upper convexity, and sometimes as a result of sheep trampling. More rarely it was due to the development of a slope foot gully. Unvegetated gully controlled slopes retreated at 8.5 mms p.a. The lower slopes of loose tipped spoil cones were found to aggrade at a rate of between 5-8 mms p.a.

The pattern of slope development discovered on the unvegetated slopes of an artificial valley developing on opencast fill was characterized by decline of the crest segment, the rapid retreat and extension of an upper element of increasing radius of curvature, parallel retreat of an abbreviating main-slope segment, the increasing differentiation of the lower slope element and rapid gully bank retreat. The mean annual rate of retreat for these slopes was 5-6 mms. The retreat of parallel vegetated slopes was much smaller at 4 mms p.a. Profile development was similar at the slope foot but the upper and mid-slopes were characterized by parallel retreat. This was also true of a slope covered by a dense seeded turf developing on fill which was part of a much larger slope system, however, this slope's retreat rate was much slower--2.3 mms p.a. A parallel but sparsely vegetated slope on the same site suffered a much greater rate of retreat: 9.5 mms p.a. and evolved by slope decline.

RUSSIAN ABSTRACT

В течение двух лет развитие склонов и холмов, развивающихся на каменноугольном холмовом отвале и на закладке, добытой открытым способом, регистрировалось при помощи специальных штифтов, предназначенных для засекания разрушении.
Линии откоса каменноугольного холмового отвала под воздействием наклона и уплотнения, за этим следовало простиранние и накопление почвы склона, параллельное отступление сокращенного отрезка склона и быстрая мелиорация (коренное улудшение) верхней выпуклости.
Склоны с растительностью отступили на 3.1 мм. (за период одного года). Это отступление наиболее ярко выразилось на южной и на западной стороне высшеупомянутых склонов. Северные и восточные стороны склона не замечали такого отступления. Склоны лишенные растительности отступили на 4.6 мм. Этот процесс обычно начинался в верхней выпуклости, а иногда следствием бараньего топота. В редких случаях это явилось результатом образования оврага с откосом почвы. Склоны с оврагами лишенными растительности отступили на 8.5 мм. Склоны, находящиеся на нижнем уровне незакрепленных отвалов почвы (с наклоном), характеризовались процессом намывания- величина эта состовляля от 5-и до 8-и мм. в год.

Структура развития, найденная на склонах лиишенных растительности и находящихся в искусственной долине с закладкой добытой открытым способом характеризовалось следующими факторами:

1. Уменьшнением вершины отрезка
2. Быстрое отступление и простирание верхней части изгиба с увеличенным радиусом
3. Параллельное отступление сокращенного отрезка склона
4. Ярко выраженное разделение нижней части склона и быстрое отступление оврага, имеющего уступ.

Средняя годовая величина отступления этих склонов составила от 5-и до 6-и мм. Отступление склонов с растительностью, находящейся на параллеле, составила всего 4 мм. в год. Развитие линии откоса у подножья склона было сходным, но в верхней и в средней (в середине) части склона наблюдалось параллельное отступление. Это же явление наблюдалось на склоне покрытом плотным семенным торфом на закладке, который принадлежал к большей группе склонов. О днако это отступление было на много медленее- 2.3 мм. в год. Параллельный склон (находящийся на параллеле) со скудным количеством растительности имел гораздо бо́льшую степень отступления, составляющую 9.5мм. в год. Это произошло в результате общего уменьшения (т.е. уменьшения размера, величины) склона.

INDEX

THE UNIVERSITY OF CHICAGO
DEPARTMENT OF GEOGRAPHY
RESEARCH PAPERS (Lithographed, 6×9 Inches)

(Available from Department of Geography, The University of Chicago, 5828 S. University Ave., Chicago, Illinois 60637. Price: $6.00 each; by series subscription, $5.00 each.)

106. SAARINEN, THOMAS F. *Perception of the Drought Hazard on the Great Plains* 1966. 183 pp.
107. SOLZMAN, DAVID M. *Waterway Industrial Sites: A Chicago Case Study* 1967. 138 pp.
108. KASPERSON, ROGER E. *The Dodecanese: Diversity and Unity in Island Politics* 1967. 184 pp.
109. LOWENTHAL, DAVID, et al. *Environmental Perception and Behavior.* 1967. 88 pp.
110. REED, WALLACE E. *Areal Interaction in India: Commodity Flows of the Bengal-Bihar Industrial Area* 1967. 210 pp.
112. BOURNE, LARRY S. *Private Redevelopment of the Central City: Spatial Processes of Structural Change in the City of Toronto* 1967. 199 pp.
113. BRUSH, JOHN E., and GAUTHIER, HOWARD L., JR. *Service Centers and Consumer Trips: Studies on the Philadelphia Metropolitan Fringe* 1968. 182 pp.
114. CLARKSON, JAMES D. *The Cultural Ecology of a Chinese Village: Cameron Highlands, Malaysia* 1968. 174 pp.
115. BURTON, IAN; KATES, ROBERT W.; and SNEAD, RODMAN E. *The Human Ecology of Coastal Flood Hazard in Megalopolis* 1968. 196 pp.
117. WONG, SHUE TUCK. *Perception of Choice and Factors Affecting Industrial Water Supply Decisions in Northeastern Illinois* 1968. 96 pp.
118. JOHNSON, DOUGLAS L. *The Nature of Nomadism* 1969. 200 pp.
119. DIENES, LESLIE. *Locational Factors and Locational Developments in the Soviet Chemical Industry* 1969. 285 pp.
120. MIHELIC, DUSAN. *The Political Element in the Port Geography of Trieste* 1969. 104 pp.
121. BAUMANN, DUANE. *The Recreational Use of Domestic Water Supply Reservoirs: Perception and Choice* 1969. 125 pp.
122. LIND, AULIS O. *Coastal Landforms of Cat Island, Bahamas: A Study of Holocene Accretionary Topography and Sea-Level Change* 1969. 156 pp.
123. WHITNEY, JOSEPH. *China: Area, Administration and Nation Building* 1970. 198 pp.
124. EARICKSON, ROBERT. *The Spatial Behavior of Hospital Patients: A Behavioral Approach to Spatial Interaction in Metropolitan Chicago* 1970. 198 pp.
125. DAY, JOHN C. *Managing the Lower Rio Grande: An Experience in International River Development* 1970. 277 pp.
126. MAC IVER, IAN. *Urban Water Supply Alternatives: Perception and Choice in the Grand Basin, Ontario* 1970. 178 pp.
127. GOHEEN, PETER G. *Victorian Toronto, 1850 to 1900: Pattern and Process of Growth* 1970. 278 pp.
128. GOOD, CHARLES M. *Rural Markets and Trade in East Africa* 1970. 252 pp.
129. MEYER, DAVID R. *Spatial Variation of Black Urban Households* 1970. 127 pp.
130. GLADFELTER, BRUCE. *Meseta and Campiña Landforms in Central Spain: A Geomorphology of the Alto Henares Basin* 1971. 204 pp.
131. NEILS, ELAINE M. *Reservation to City: Indian Urbanization and Federal Relocation* 1971. 200 pp.
132. MOLINE, NORMAN T. *Mobility and the Small Town, 1900–1930* 1971. 169 pp.
133. SCHWIND, PAUL J. *Migration and Regional Development in the United States, 1950–1960* 1971. 170 pp.
134. PYLE, GERALD F. *Heart Disease, Cancer and Stroke in Chicago: A Geographical Analysis with Facilities Plans for 1980* 1971. 292 pp.
135. JOHNSON, JAMES F. *Renovated Waste Water: An Alternative Source of Municipal Water Supply in the U.S.* 1971. 155 pp.
136. BUTZER, KARL W. *Recent History of an Ethiopian Delta: The Omo River and the Level of Lake Rudolf* 1971. 184 pp.
137. HARRIS, CHAUNCY D. *Annotated World List of Selected Current Geographical Serials in English, French, and German* 3rd edition 1971. 77 pp.
138. HARRIS, CHAUNCY D., and FELLMANN, JEROME D. *International List of Geographical Serials* 2nd edition 1971. 267 pp.
139. MC MANIS, DOUGLAS R. *European Impressions of the New England Coast, 1497–1620* 1972. 147 pp.
140. COHEN, YEHOSHUA S. *Diffusion of an Innovation in an Urban System: The Spread of Planned Regional Shopping Centers in the United States, 1949–1968* 1972. 136 pp.

141. MITCHELL, NORA. *The Indian Hill-Station: Kodaikanal* 1972. 199 pp.

142. PLATT, RUTHERFORD H. *The Open Space Decision Process: Spatial Allocation of Costs and Benefits* 1972. 189 pp.

143. GOLANT, STEPHEN M. *The Residential Location and Spatial Behavior of the Elderly: A Canadian Example* 1972. 226 pp.

144. PANNELL, CLIFTON W. *T'ai-chung, T'ai-wan: Structure and Function* 1973. 200 pp.

145. LANKFORD, PHILIP M. *Regional Incomes in the United States, 1929–1967: Level, Distribution, Stability, and Growth* 1972. 137 pp.

146. FREEMAN, DONALD B. *International Trade, Migration, and Capital Flows: A Quantitative Analysis of Spatial Economic Interaction* 1973. 202 pp.

147. MYERS, SARAH K. *Language Shift Among Migrants to Lima, Peru* 1973. 204 pp.

148. JOHNSON, DOUGLAS L. *Jabal al-Akhdar, Cyrenaica: An Historical Geography of Settlement and Livelihood* 1973. 240 pp.

149. YEUNG, YUE-MAN. *National Development Policy and Urban Transformation in Singapore: A Study of Public Housing and the Marketing System* 1973. 204 pp.

150. HALL, FRED L. *Location Criteria for High Schools: Student Transportation and Racial Integration* 1973. 156 pp.

151. ROSENBERG, TERRY J. *Residence, Employment, and Mobility of Puerto Ricans in New York City* 1974. 230 pp.

152. MIKESELL, MARVIN W., editor. *Geographers Abroad: Essays on the Problems and Prospects of Research in Foreign Areas* 1973. 296 pp.

153. OSBORN, JAMES. *Area, Development Policy, and the Middle City in Malaysia* 1974. 273 pp.

154. WACHT, WALTER F. *The Domestic Air Transportation Network of the United States* 1974. 98 pp.

155. BERRY, BRIAN J. L., et al. *Land Use, Urban Form and Environmental Quality* 1974. 464 pp.

156. MITCHELL, JAMES K. *Community Response to Coastal Erosion: Individual and Collective Adjustments to Hazard on the Atlantic Shore* 1974. 209 pp.

157. COOK, GILLIAN P. *Spatial Dynamics of Business Growth in the Witwatersrand* 1975. 143 pp.

158. STARR, JOHN T., JR. *The Evolution of Unit Train Operations in the United States: 1960–1969—A Decade of Experience* 1976. 247 pp.

159. PYLE, GERALD F. *The Spatial Dynamics of Crime* 1974. 220 pp.

160. MEYER, JUDITH W. *Diffusion of an American Montessori Education* 1975. 109 pp.

161. SCHMID, JAMES A. *Urban Vegetation: A Review and Chicago Case Study* 1975. 280 pp.

162. LAMB, RICHARD. *Metropolitan Impacts on Rural America* 1975. 210 pp.

163. FEDOR, THOMAS. *Patterns of Urban Growth in the Russian Empire during the Nineteenth Century* 1975. 275 pp.

164. HARRIS, CHAUNCY D. *Guide to Geographical Bibliographies and Reference Works in Russian or on the Soviet Union* 1975. 496 pp.

165. JONES, DONALD W. *Migration and Urban Unemployment in Dualistic Economic Development* 1975. 186 pp.

166. BEDNARZ, ROBERT S. *The Effect of Air Pollution on Property Value in Chicago* 1975. 118 pp.

167. HANNEMANN, MANFRED. *The Diffusion of the Reformation in Southwestern Germany, 1518-1534* 1975. 248 pp.

168. SUBLETT, MICHAEL D. *Farmers on the Road. Interfarm Migration and the Farming of Noncontiguous Lands in Three Midwestern Townships, 1939-1969* 1975. 228 pp.

169. STETZER, DONALD FOSTER. *Special Districts in Cook County: Toward a Geography of Local Government* 1975. 189 pp.

170. EARLE, CARVILLE V. *The Evolution of a Tidewater Settlement System: All Hallow's Parish, Maryland, 1650–1783* 1975. 249 pp.

171. SPODEK, HOWARD. *Urban-Rural Integration in Regional Development: A Case Study of Saurashtra, India—1800–1960* 1976. 156 pp.

172. COHEN, YEHOSHUA S. and BERRY, BRIAN J. L. *Spatial Components of Manufacturing Change* 1975. 272 pp.

173. HAYES, CHARLES R. *The Dispersed City: The Case of Piedmont, North Carolina* 1976. 169 pp.

174. CARGO, DOUGLAS B. *Solid Wastes: Factors Influencing Generation Rates* 1977.

175. GILLARD, QUENTIN. *Incomes and Accessibility. Metropolitan Labor Force Participation, Commuting, and Income Differentials in the United States, 1960–1970* 1977. 140 pp.

176. MORGAN, DAVID J. *Patterns of Population Distribution: A Residential Preference Model and Its Dynamic* 1977.

177. STOKES, HOUSTON H.; JONES, DONALD W. and NEUBURGER, HUGH M. *Unemployment and Adjustment in the Labor Market: A Comparison between the Regional and National Responses* 1975. 135 pp.

178. PICCAGLI, GIORGIO ANTONIO. *Racial Transition in Chicago Public Schools. An Examination of the Tipping Point Hypothesis, 1963–1971* 1977.

179. HARRIS, CHAUNCY D. *Bibliography of Geography. Part I. Introduction to General Aids* 1976. 288 pp.

180. CARR, CLAUDIA J. *Pastoralism in Crisis. The Dasanetch and their Ethiopian Lands.* 1977. 339 pp.

181. GOODWIN, GARY C. *Cherokees in Transition: A Study of Changing Culture and Environment Prior to 1775.* 1977. 221 pp.

182. KNIGHT, DAVID B. *A Capital for Canada: Conflict and Compromise in the Nineteenth Century.* 1977. 359 pp.

183. HAIGH, MARTIN J. *The Evolution of Slopes on Artificial Landforms: Blaenavon, Gwent.* 1978. 311 pp.

184. FINK, L. DEE. *Listening to the Learner. An Exploratory Study of Personal Meaning in College Geography Courses.* 1977. 200 pp.

185. HELGREN, DAVID M. *Rivers of Diamonds: An Alluvial History of the Lower Vaal Basin.* 1978.

186. BUTZER, KARL W., editor. *Dimensions of Human Geography: Essays on Some Familiar and Neglected Themes.* 1978. 201 pp.